JOURNAL OF GREEN ENGINEERING

Volume 4, No. 3 (April 2014)

JOURNAL OF GREEN ENGINEERING

Aims and Scopes

Journal of Green Engineering will publish original, high quality, peer-reviewed research papers and review articles dealing with environmentally safe engineering including their systems. Paper submission is solicited on:

- Theoretical and numerical modeling of environmentally safe electrical engineering devices and systems.
- Simulation of performance of innovative energy supply systems including renewable energy systems, as well as energy harvesting systems.
- Modeling and optimization of human environmentally conscientiousness environment (especially related to electromagnetics and acoustics).
- Modeling and optimization of applications of engineering sciences and technology to medicine and biology.
- Advances in modeling including optimization, product modeling, fault detection and diagnostics, inverse models.
- Advances in software and systems interoperability, validation and calibration techniques. Simulation tools for sustainable environment (especially electromagnetic, and acoustic).
- Experiences on teaching environmentally safe engineering (including applications of engineering sciences and technology to medicine and biology).

All these topics may be addressed from a global scale to a microscopic scale, and for different phases during the life cycle.

Published, sold and distributed by:
River Publishers
Niels Jernes Vej 10
9220 Aalborg Ø
Denmark

Tel.: +45369953197
www.riverpublishers.com

Journal of Green Engineering is published four times a year.
Publication programme, 2013–2014: Volume 4 (4 issues)

ISSN 1904-4720 (Print Version)
ISSN 2245-4586 (Online Version)
ISBN 978-87-93237-58-2

JOURNAL OF GREEN ENGINEERING

Volume 4 No. 3 April 2014

Performance Analysis of Inter and Intra IMS Network

A Bagubali,[1] Ankit Agarwal,[2] V Prithiviraj[3] and P. S Mallick[4]

[1,2,4] *Vellore Institute of Technology University, Vellore-632014, Tamilnadu, India*
bagubali@vit.ac.in; ankit.agrwl80@gmail.com; psmallick@vit.ac.in
[3]*Rajalakshmi Institute of Technology, Chennai, Tamilnadu, India*
profvpraj@gmail.com

Received 19 August 2014; Accepted 19 November 2014;
Publication 19 March 2015

Abstract

The IP Multimedia Subsystem (IMS) is a network architecture that consists of different network elements linked by standard interfaces. It facilitates the delivery of multimedia services based on Internet Protocol (IP) and makes the integration of services with the internet much easier. It is a method to achieve the convergence of fixed and mobile communication devices. Optimized transmission of voice, data and video communications among the users can be achieved using IMS architecture, independent of the user's location and devices. IMS enables the user to use different services in a more efficient way in terms of energy saving. Using this technology, it is possible to make the network distribution in a better way. It reduces the energy consumption because of the capability to switch network whenever possible. This paper focuses on establishment of end to end session between two IMS terminals and analysing the performance for different applications. The performance is measured on the basis of traffic parameters and call setup delay. We have compared Inter and Intra IMS network based on a model that ensures the delivery of services without any data loss and with minimum delay.

Keywords: IMS, SIP, HSS, VoIP, QoS.

Journal of Green Engineering, Vol. 4, 175–194.
doi: 10.13052/jge1904-4720.431

1 Introduction

The world of communication is emerging towards the trend that supports integration of user with the outside world in a much easier way. IMS solves the continuing demand of interoperability where the user can enjoy wide range of services despite their location and access medium. Interoperability is achieved by the convergence of fixed and mobile communication devices and networks. Fixed-mobile convergence means a device can change its connection from one network to another network, say wired to wireless network. While using IMS at both the terminals, it is easier for the users to adopt any network without any interrupt in the services. The application servers used in IMS, performs the management operations, that basically takes care of energy consumption. It has become the biggest challenge in today's generation to achieve less energy consumptive method while using multimedia. IMS mainly uses Session Initiation Protocol (SIP), as the protocol for unifying the applications. It is responsible for VoIP call setup and handling [1]. A standard IMS architecture comprises of many SIP servers and Home Subscriber Server(HSS) also known as user database [2]. HSS contains information about the user subscription and provides subscriber's location and IP information. The purpose of SIP is to initiate the session among user agents. The proxy servers used in IMS are the major part its architecture. They are collectively given the name as Call Session Control Functions (CSCF). These functions can be further classified as Proxy-CSCF, Serving-CSCF, and Interrogating-CSCF [3]. P-CSCF is a starting point of contact with which the user initiates to send the call request. It provides authentication to an user and inspects the signal ensuring that thc IMS terminal do not misbehave. S-CSCF is a SIP server responsible for registrations by verifying the address assigned to an user. It takes a decision upon choosing the application server to which the SIP message will be forwarded. I-CSCF is responsible for contacting HSS to take the user location and forwarding the corresponding response to S-CSCF.

The rest of this paper has been designed as follows: Next subsections describe about the architecture and different processes involved related to IMS. Section 2 discusses the related work that has been done in past. The model that has been used in this paper, is described in Section 3. Simulation results have been presented in Section 4 and finally the conclusions are made in Section 5.

1.1 IMS Architecture

The IMS architecture includes three separated layers, each functioning independently. The three layers are named as:

- Transport and End Point Layer: It is responsible for changing the format of incoming media from analog\digital to real time transfer protocol and SIP protocol. This layer separates the access layers from the IP network above it. It is responsible for assigning the IP address and registering of devices with the upper layers.
- Control and Session Layer: This layer forms the logical connections between various network elements. As suggested by the name itself, it controls the authentication, routing and traffic between transport and service layer. It also provides an interface between service layer and other services. Layer consists of two important IMS network elements i.e. CSCF and HSS. CSCF manages the interaction among the users with the help of SIP servers. HSS contains the whole information about the user subscription which helps in authenticating the user with the network.
- Application Services Layer: The application services layer consists of several Application Servers, responsible for performing different functions on user's session. This layer is responsible for deployment of new services in the IMS network with the existing services.

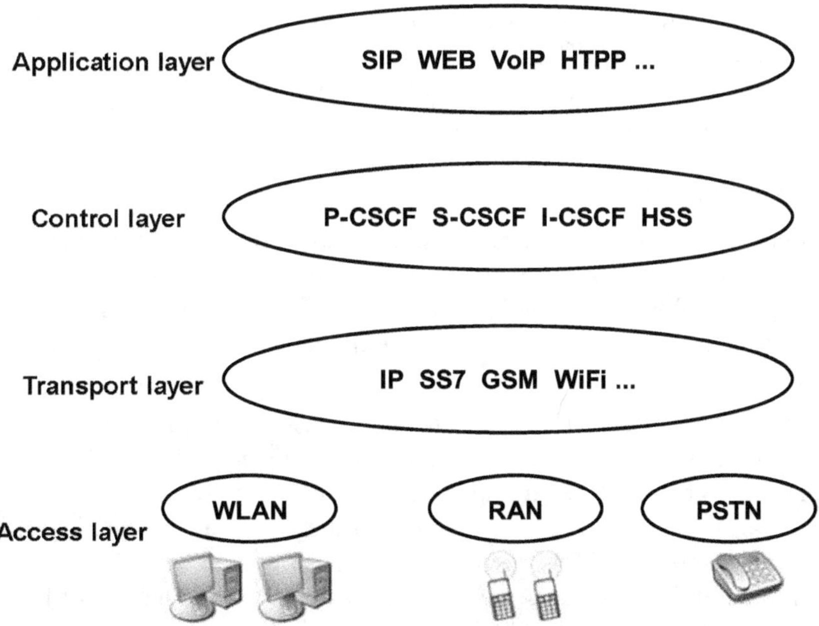

Figure 1 IMS three layered architecture

1.2 Need of IMS

As stated earlier, IMS provides an integrated architecture for the transmission of different multimedia services over internet protocol. The main motive behind using the IMS architecture is to manage the network in a much easier way. IMS provides the opportunity to network operators for delivering more reliable and profitable multimedia services with the existing services. It provides the facility of migration from circuit switching services to packet switching domain. Different users can be connected as IMS terminal and the initiated call can be easily handed off from one network to another network without any interruption. IMS architecture is access independent therefore the services are delivered independent of access technology [4]. It supports for roaming between different networks. IMS has provided the options to combine ongoing voice call with different multimedia elements like sharing photos and videos while talking. IMS enables the single integrated network for all the access types. Therefore the cost has come down for the service providers. The IMS compatible systems are designed to support multiple application servers which means that same architecture can be reused for new services that is made to focus on actual service to be provided. The authentication process is more simpler and has been standardized in IMS as compared to non-IMS architecture. Once the subscriber is authenticated through IMS, it can access all the IMS services. In non-IMS scenario each service has its own way to authenticate the user. Conclusively, IMS provides a better secure connection with the outside world giving the facility to change the communication mode smoothly.

1.3 IMS Challenges

IMS is a new emerging technology towards the evolution of user integration with the internet. There are several challenges and objectives that can be addressed over deployment of IMS architecture. IMS has emerged as a solution to manage the network smoothly. Some of the challenges are mentioned below related to the IMS architecture

- Quality Of Service (QOS): One of the major issues in implementing IMS architecture is to enhance the Quality of service parameters, so that user can meet its demands regarding bandwidth and delay requirements. It is essential to provide secure connection to the user that protects it from creating a traffic load for other users.
- Multiple Services: As stated earlier, one of the important feature of IMS is to enable different services simultaneously. IMS is mainly designed to achieve this purpose so that user can be able to enjoy wide range of

services at a time. Handling of multiple services without any delay and interrupt during the call flow has come up as a challenge before IMS. Another issue is regarding the security of these services, IMS should be able to provide a secure connection for each of the services.

- Authentication: IMS architecture uses a completely different methodology to provide authentication to a set of users. Authentication process needs to be strong in terms of security so that private data cannot be reached to malicious agents or unethical hackers. The registration request to IMS terminal has to be properly examined by the protocols used so that unidentified user can be easily ignored.
- Energy Saving: The major issue in today's world is related to energy. Main challenge is to develop a technology considering the environmental requirements. During the wireless transmission of any media, a lot of energy is consumed in terms of battery power. IMS provides the best network switching mechanism which enables it to reduce energy consumption whenever required.

1.4 IMS Registration and Session Establishment

In order to initiate a call among the users, IMS architecture must know where the user(User Equipment(UE)) is located. All the users must register to the network during the activation. If the user changes its location it must re-register to the network so that updated location can be retrieved. During the registration process, the request is sent to P-CSCF, which is the initial point to make contact with another user. On receiving the request, this SIP proxy needs to locate I-CSCF to know about the user's location and other information. I-CSCF sends a query to HSS in order to get the subscription information about the user. Once the registration request reaches to S-CSCF, it challenges the user by sending an Unauthorised response, so that it can receive the authorisation data from the user. This data is matched with the data available with HSS, and if verified, the positive authentication response is reached back to the user.

Once the registration process gets completed, if anyone wants to establish a session with user at the other end, it sends a SIP invite message to the first point of contact i.e. P-CSCF. After receiving invite message, it checks for the location of called user, whether it belongs to the same network or not. If network is found to be same for both calling and called party, then P-CSCF directly contacts to the assigned S-CSCF for called user. Invite message is sent to the S-CSCF, and now S-CSCF will be responsible for the arrival of invite message to the final receiver. But, if the called party belongs to other IMS

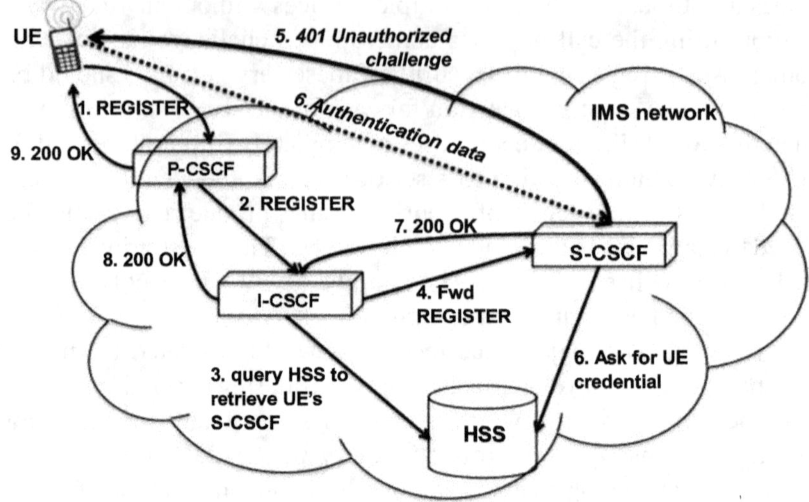

Figure 2 IMS registration

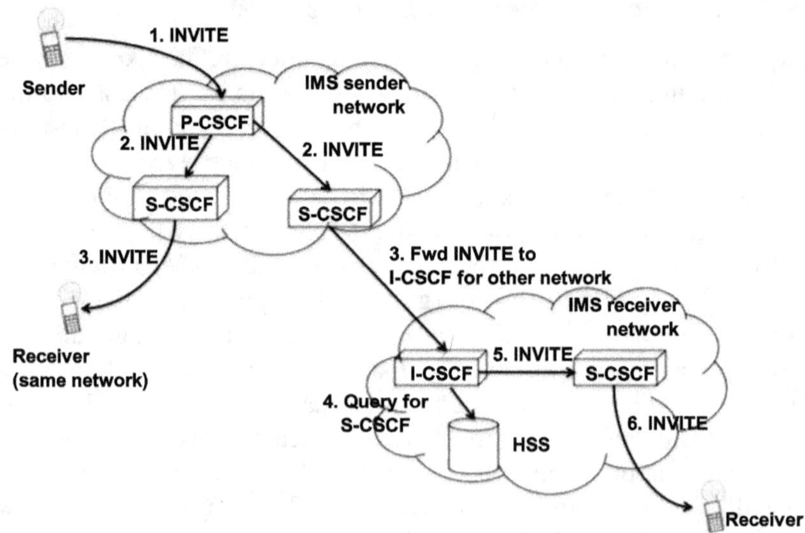

Figure 3 IMS session establishment

network, then P-CSCF has to locate corresponding I-CSCF of the external network, so that location of called party can be retrieved. I-CSCF contacts to HSS that may contain the user's information. If the user has not registered to the network or belong to some other network then HSS returns NOT FOUND response to the query. In case when the user's location is found in HSS, the Invite message is sent to S-CSCF and finally delivered to the receiver. The receiver then sends an acknowledgement to caller party, and at this point SIP session is successfully established [5].

2 Related Work

Most of the previous work related to IMS has focussed mainly on protocol development and complexity of SIP. However, to set up an uninterrupted connection it is important to take into account, the call setup delay. This analysis of delay parameter will be able to predict the data loss during the hand off process from one network to another network. Research of Eyers et al [6] was based on SIP call setup delay. It mainly focuses on data loss due to UDP. Kist et al [7] presented the signalling delays, assuming the queuing delay to be less than 5ms. V.S. Abhayawardhana, R. Babbage [8] presented a traffic model for the IP multimedia subsystem. In this model, signalling traffic was created to and from HSS in three processes i.e. registration, session setup and subscription. In [9], the authors have analyzed the SIP IMS network based on signalling delays, and their analysis was based on real time server traffic. Other researches has mainly concentrated on network performance evaluation based on different network parameters [10–12]. M. Melnyk and A. Jukan [13], studied the call setup efficiency in IP wireless networks. It presented the session forming process in IMS, where both the source and destination node are using same technology standards.

3 Description of the Model

To evaluate the performance of IMS network model and to solve the challenges related to it, we have designed and tested the network model in OPNET modeler 14.0. The network is designed in such a way that all the intermediate proxy servers have been used to complete the process of session establishment among the users. Two IMS networks are established in different regions, and three users are made to register with each IMS terminal. The designed network basically consists of SIP proxy servers at each end, connected with the Ethernet routers. As mentioned earlier, intermediate proxy servers are P-CSCF, S-CSCF, I-CSCF, each playing different roles in establishing the session among two users. Since, it's a internetwork scenario, we have used IP cloud to make a connection between the two networks. Traffic has been created by

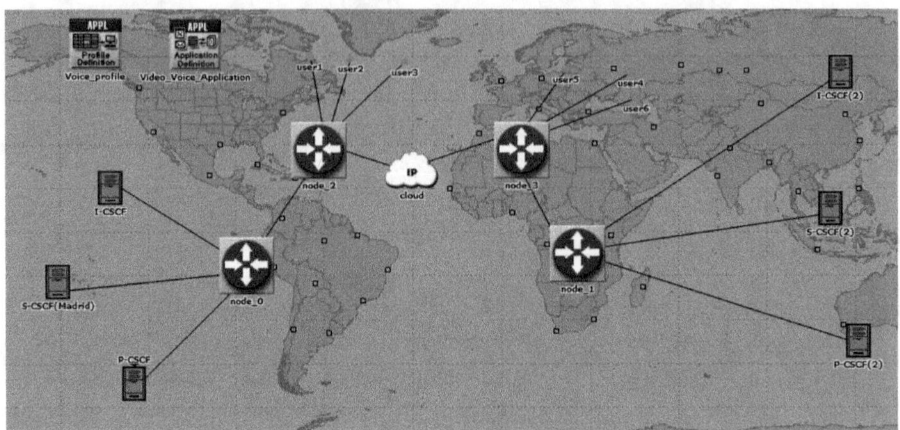

Figure 4 IMS internetwork scenario

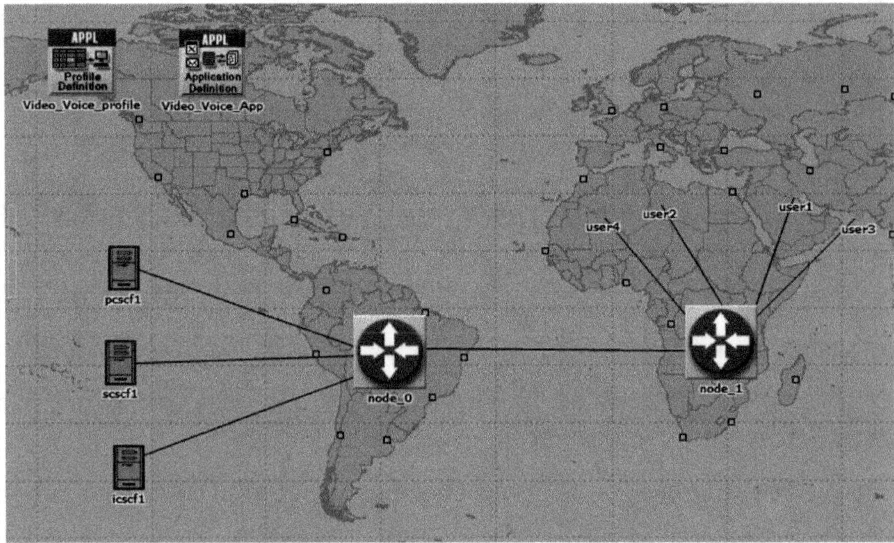

Figure 5 IMS intranetwork scenario

making use of different applications with VoIP. In this model, we have used Video Conferencing as secondary application which basically creates traffic while transmitting voice packets from one end to another.

If any user wishes to make a contact with the user located in same network, then the intermediate server I-CSCF will directly send invite request to the corresponding S-CSCF. There will be no need to send a query to HSS for

Table 1 Applications

Applications	Attributes	
Voice Application	Encoder Scheme	PCM quality speech
	Voice Frames/packet	5
	Compression Delay	0.02 sec
Video Conferencing	Type	Low resolution video
	Frame Interarrival	10 frames/sec
	Frame size	128x120 pixels

the user location as the called user belongs to same network. To test the performance of session establishment within the same network, we have created another scenario containing two users registered to the same IMS network. All the attributes and applications are kept same as previous scenario, the only change that has been made is, only one IMS network is present instead of two. Since, it's a intranetwork scenario, there is no need to use IP cloud for the connection setup. Only Ethernet routers have been used to connect the proxy servers and the users.

3.1 Application Services

In order to generate the traffic among the nodes, two simultaneous applications have been used in this model. One is voice application and the another application is video conferencing. Starting time of execution for both the applications have been given same, so that motive of designing IMS network can be successfully achieved.

3.2 Mathematical Delay Model

This section analyses the delay occurred at different stages for the designed IMS networks. IMS delay collectively comprises of three types of delay named as, processing delay, transmission delay and queuing delay.

$$T = t_p + t_t + t_q \tag{1}$$

where T, t_p, t_t and t_q represents total average IMS delay, processing delay, transmission delay and queing delay.

- Transmission delay: It is the delay occurred while transmitting signal message from UE to P-CSCF terminal From [13], it is assumed that there are total 8 messages included in the registration process from UE to P-CSCF. The IMS registration transmission delay can be written as:

$$t_t(ims) = 8 \times t_{tcp} \quad (2)$$

- Processing delay: It is the delay due to packet wrapping and unwrapping in the network layer. The IMS processing delay can be written as:

$$\begin{aligned} t_p(ims) = {} & 4 \times t_{p\,-\,UE} + 10 \times t_{p\,-\,PCSCF} + 6 \times t_{p\,-\,ICSCF} \\ & + 4 \times t_{p\,-\,HSS} + 8 \times t_{p\,-\,SCSCF} \end{aligned} \quad (3)$$

Here, $t_{p\text{-}UE}$, $t_{p\text{-}PCSCF}$, $t_{p\text{-}ICSCF}$, $t_{p\text{-}HSS}$, $t_{p\text{-}SCSCF}$ represents the delay incurred for message processing at the respective nodes.

- Queuing delay: It is the delay occurred, because of queuing of packets at the network nodes. The queuing model is assumed to be of type M/M/1 queuing model, in which the arrivals occurs at rate λ according to a poisson process and a single server serves the customers one at a time from the front. The IMS queuing delay can be written as:

$$\begin{aligned} t_q(ims) = {} & 4 \times E(t_{q\,-\,UE}) + 10 \times E(t_{q\,-\,PCSCF}) \\ & + 6 \times E(t_{q\,-\,ICSCF}) + 4 \times E(t_{q\,-\,HSS}) \\ & + 8 \times E(t_{q\,-\,SCSCF}) \end{aligned} \quad (4)$$

Here, $E(t_{q\text{-}UE})$, $E(t_{q\text{-}PCSCF})$, $E(t_{q\text{-}ICSCF})$, $E(t_{q\text{-}HSS})$, $E(t_{q\text{-}SCSCF})$ represents the expected unit packet queing delay at the respective nodes.

As a combination of all the three delays, the total IMS registration delay can be written as:

$$\textit{Total IMS delay} = t_t(ims) + t_p(ims) + t_q(ims) \quad (5)$$

4 Simulation Results

4.1 SIP Call Set UP Time

Figure 6 analyzes the SIP call set up time for Inter/Intra IMS networks. It is observed that time taken in Intranetwork scenario is very less as compared to time taken in Internetwork scenario. This is due to the fact that in Intranetwork scenario both the users are located within the same network, that is why time taken to process the SIP call setup request is very less. While in Internetwork scenario users belong to different networks, therefore time taken is more, as the location of the called user has to be retrieved first and then the request is being processed.

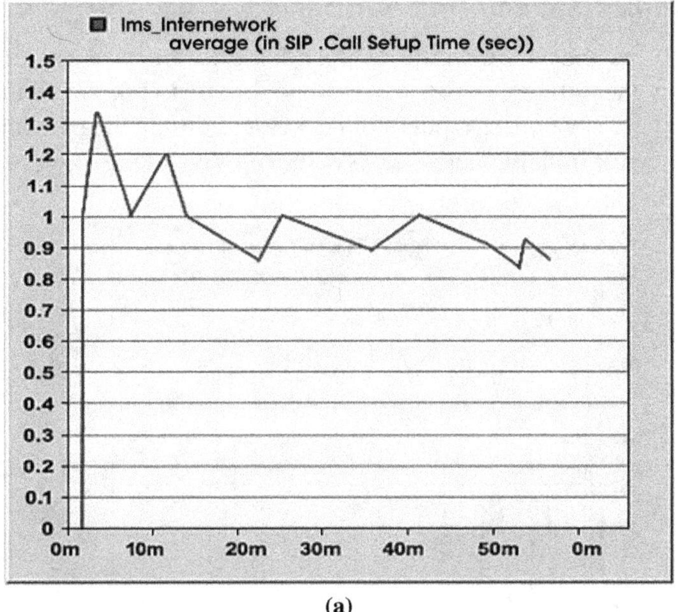

(a)

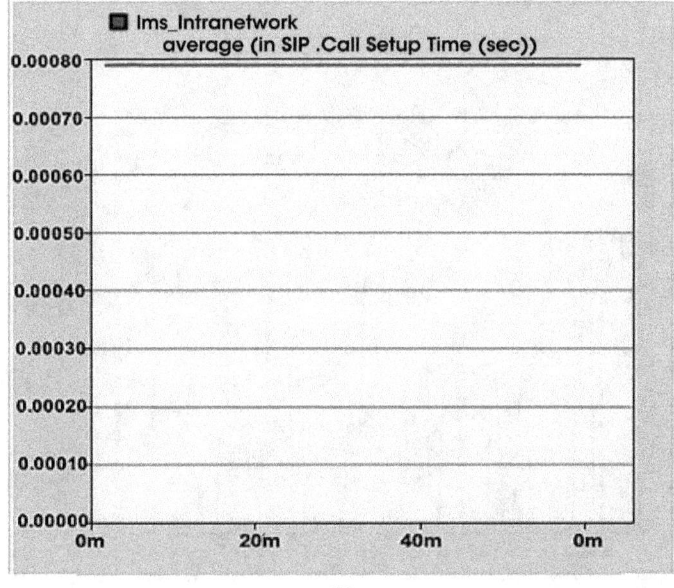

(b)

Figure 6 SIP call setup time (a) Internetwork (b) Intranetwork

4.2 Voice Traffic Sent and Traffic Received

Figure 7 analyzes the voice traffic sent and traffic received in terms of packets/sec for the Intranetwork scenario and Internetwork scenario. It is observed that there is very less amount of packet loss during the transmission. So the efficiency for both the networks is achieved very high.

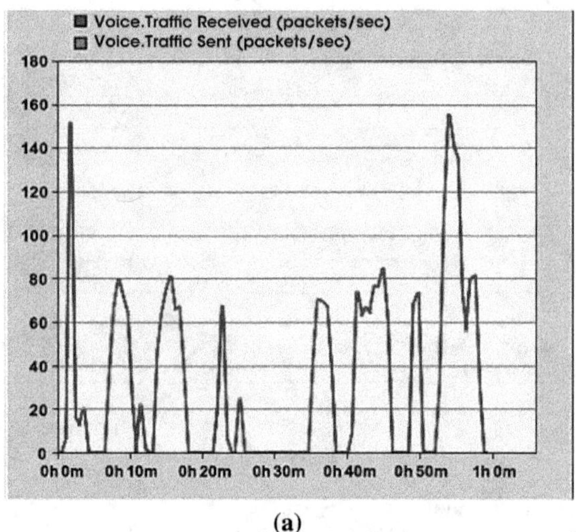

(a)

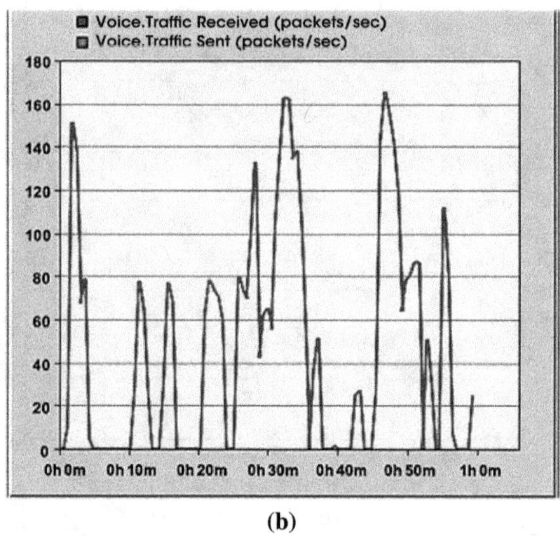

(b)

Figure 7 Voice Traffic Sent and Traffic Received (a) Internetwork (b) Intranetwork

4.3 Video Traffic Sent and Traffic Received

Figure 8 analyzes the traffic sent and received for the video conferencing application in Intra/Inter IMS network. It is again observed that packet loss in both the networks is negligible. Therefore, it can be concluded that whenever the IMS network is used, simultaneous multimedia application can be transmitted/received without any data loss.

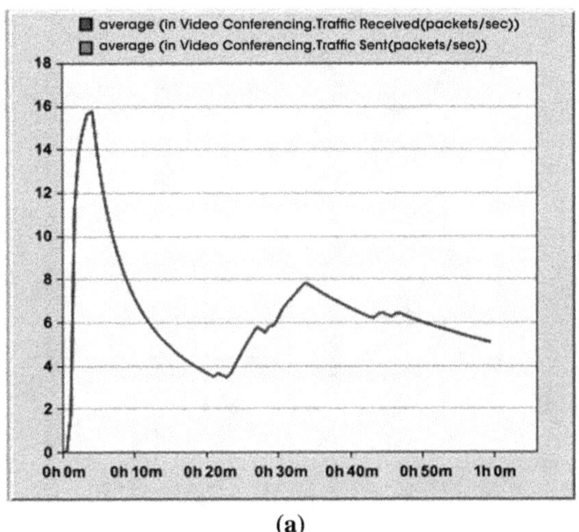

(a)

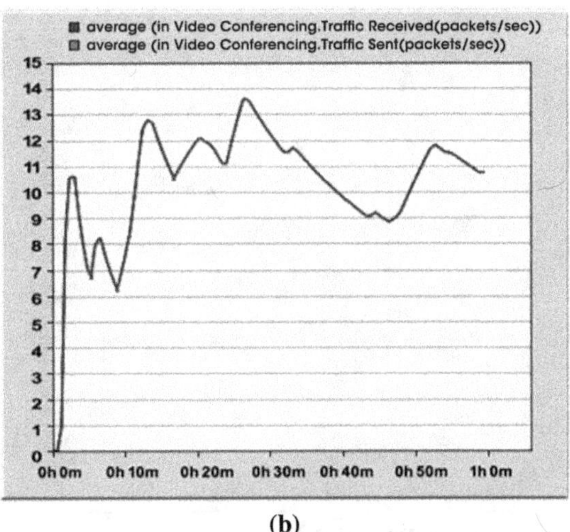

(b)

Figure 8 Video Traffic Sent and Traffic Received (a) Internetwork (b) Intranetwork

4.4 SIP UAC Active Calls and Call Duration

Figure 9(a) compares the total number of UAC active calls for both the Inter/Intra network scenario. These are the calls which are activated form user1 to user2 after the SIP request is accepted at the terminal.

Figure 9(b) compares the active call duration for both the scenarios. Call duration is more in case of internetwork scenario as compared to intranetwork.

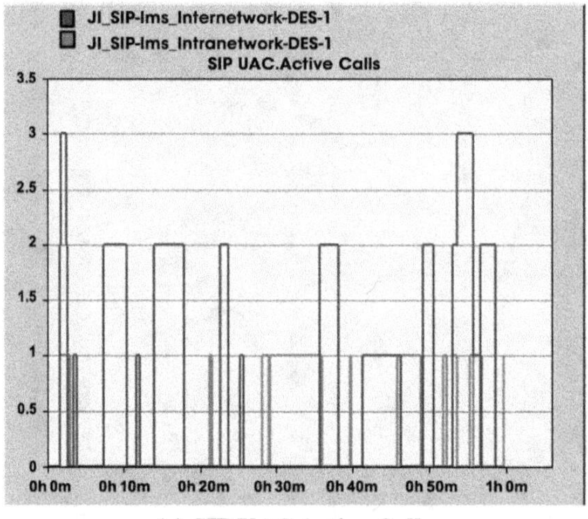

(a) SIP UAC Active Calls

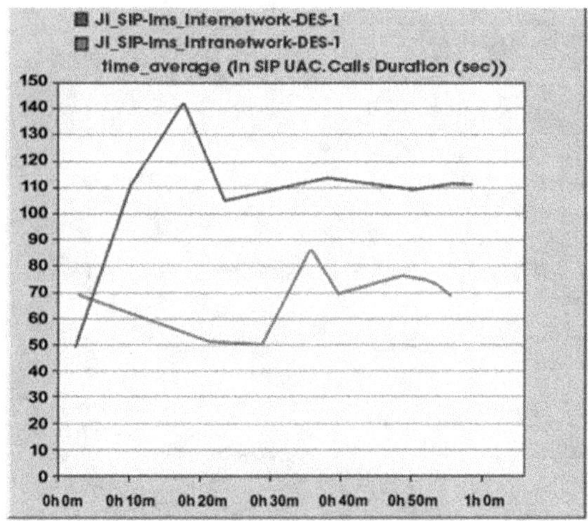

(b) ISP UAC Call Duration

Figure 9 SIP Call Information

5 Conclusion

As already discussed in this paper, use of IMS network makes the hand off process much easier during the call transfer from one network to another. We have analyzed the performance of IMS network based on the call setup delay and traffic transferred from one user to another. In this paper IMS network has been designed in two phases .i.e. Intranetwork and Internetwork. We have simulated both the scenarios using OPNET modeler tool, that takes into account all the intermediate servers like S-CSCF, P-CSCF and I-CSCF. It has been concluded that time taken to setup a call is very less in intranetwork as compared to internetwork. Also we have seen that there is no data loss in both the networks for voice as well as video transfer. The use of IMS leads to a solution of energy consumption during wireless transmissions. It has become possible because protocols adopted by IMS mainly consider the node according to energy consumption and also handoff techniques are appropriately adopted by IMS to achieve the goal of energy saving wireless environment. The network switching mechanism in IMS has been designed in such a way that it always prefers to choose the network having high range with less energy consumption.

For future work, we are planning to integrate this IMS network model with other networks, to achieve the target of proper hand off while switching between different networks during a call. This IMS model may act as intermediate network, so that call can be transferred without any interruption.

References

[1] N. Rajagopal, M. Devetsikiotis, "Modeling and Optimization for the Design of IMS Networks", 39th Annual Simulation Symposium 2006, pp: 34–41.

[2] P. Agrawal, et. al, "IP Multimedia Subsystems in 3GPP I\& 3GPP2:Overview and Scalability issues", IEEE Communications Magazine, January, 2008.

[3] Poikselka, M.; Mayer, G.; Khartabil, H.;Niemi, A. The IMS: IP Multimedia Concepts and Services in the Mobile Domain. England: WILEY, 448 p. ISBN-10 0-470-87113-X, 2004

[4] Camarillo, G.; García-Martín, M. A(2008). The 3G IP Multimedia Subsystem (IMS) Merging the Internet and the Cellular Worlds. Third Edition. England: WILEY, 618 p. ISBN-13978-0-470-51662-1

[5] 3GPP IP Multimedia Subsystem (IMS), 3GPP TS 23.228, 2013

[6] T. Eyers and H. Schulzrinne. Predicting internet telephony call setup delay In IPTel 2000, First IP Telephony Workshop, Berlin, Germany, 2000.
[7] A. Kist and R. Harris. SIP signalling delay in 3GPP. In Proceedings of Sixth International Symposium on Communications Interworking of IFIP-Interworking 2002, Perth, Australia, October 13–16 2002.
[8] V.S. Abhayawardhana, R. Babbage, "A Traffic Model for the IP Multimedia Subsystem (IMS)", IEEE 65th Vehicular Technology Conference, 22–25 April 2007, pp. 783–787.
[9] J.-S. Wu and P.-Y. Wang. The performance analysis of SIPT signaling system in carrier class VoIP network. (AINA'03).
[10] S. Pandey, V. Jain, D. Das, V. Planat, R. Periannan, "Performance Study of IMS Signaling Plane", IP Multimedia Subsystem Architecture and Applications, International Conference, 6–8 Dec. 2007, pp.1–5.
[11] A. Munir, "Analysis of SIP-Based IMS Session Establishment Signaling for WiMax-3G Networks", Networking and Services, 4th International Conference, 16–21 March 2008, pp. 282–287.
[12] A. A Kist, RJ. Hark, "SIP signalling delay in 3GPP", IEEE 6th international Symposium on Communications Interworking of IFIP interworking, Fremantle WA, October 13–16, 2002.
[13] M. Melnyk and A. Jukan. On Signaling Efficiency for Call Setup in all-IP Wireless Networks. In Proceedings of IEEE International Conference on Communication (ICC'06), Istanbul, Turkey, June 2006.

Biographies

A. Bagubali obtained his B.Tech in electronics and communication engineering from VIT University, Vellore, in 2005, and M.Tech in communication engineering from VIT University, Vellore, in 2007. Currently he is senior assistant professor in VIT University, Vellore. His research interests include IP multimedia subsystem applied to 3G and 4G networks.

A. Agarwal obtained his B.Tech degree in Electronics and Communications Engineering from VIT University, Vellore, in 2014. Currently he is working as software Engineer. His area of interests include wireless networking.

Dr. V. Prithviraj is now the principal of Rajalakshmi Institute of Technology, Chennai. He had served from August 2008–January 2013 as the Principal of Pondicherry Engineering College. He completed his Bachelor of Engineering (Electronics and Communication Engineering) in 1972 from the College of Engineering, Guindy Madras University, his M.S. by research in 1982 from IIT, Madras, and his Ph.D in 1999 from IIT, Kharagpur (research area–Signal Processing Techniques in Array Antennas Systems). He is one of the founding members of Pondicherry Engineering College and served the institute from July 1985–2013. He was holding the position of Dean-in-charge, School of Engineering, Pondicherry University from May 2009–January 2013.

He has been teaching for over 30 years and held the position of Head of the Electronics and Communications Engineering Department at PEC from June 2003–2006. He has published over 80 technical research papers. He has also held the position of Director, IT for Government of Pondicherry (2002–2005). Currently, he is a Member of Expert Committee for monitoring International Indo-French projects in the field of Information Technology as well as Regional Committee of AICTE for Tamil Nadu and Pondicherry. He has a keen interest in research and development projects and provided leadership in many successful projects sponsored by various organizations such as DRDO, ISRO, Department of Electronics and Department of Information Technology at IITs and PEC. He is the recipient of the IEEE International Student Branch Award in 1984 and the EDI Award for best technical paper, entitled "COFDM for Telemedicine Applications", in 2007. He is a Life Member of ISTE and Member of EMC Engineers and IEEE USA. His areas of interest include Broadband and Wireless Communication, Mobile Computing, VLSI for Wireless Applications, Tele Medicine, SDR, Cognitive Radio and e-Governance Applications.

P. S. Mallick, received Ph.D from Jadavpur University, Kolkata, India. He worked 4.5 years in a Sweden based electronics industry named *IAAB Electronics* as a Technical Head. He has 12 years of Teaching experience where he led various research teams and developed "Online Lab in Microelectronics", "Monte Carlo Simulator for Compound Semiconductors", "Nanostructured MIM Capacitor" and "Low cost Electric Fencer". His current area of research interest includes Nanoscale CMOS, Nanoelectronics and VLSI Engineering. He has published 48 research papers in different Journals and Conferences of International repute and authored a book on *Matlab and Simulink.* At Present Dr Mallick is working for School of Electrical Engineering, VIT University, Tamilnadu, India, as a Professor and Dean. Dr Mallick has received the prestigious *Jawaharlal Nehru Scholarship* in 1998 for his doctoral research work. He is one of the enlisted technical innovators of India in 2007. He is a Sr. member of IEEE, IEEE-EDS, Life member of IACS, ISTE, and Indian Laser Association. He has successfully completed six research projects funded by the Govt. of India and organized several International Conferences.

TCP/IP Based Vehicle Tracking and Fuel Monitoring Using Low Power Microcontroller

M. Shanmugasundaram, D. Karthikeyan, K. Arul Prasath and R. Sri Raghav

Vellore Institute of Technology University, Vellore-632014, Tamilnadu, India
Correspoding author: mshanmugasundaram@vit.ac.in

Received 4 August 2014; Accepted 13 November 2014; Publication 19 March 2015

Abstract

This paper is for transportation security and fleet management. An embedded device is designed and fabricated into PCB. The embedded device having SIM908, integrated GSM/GPS module, is kept inside vehicle which acquires the speed and location of the vehicle and send data to the server. The fuel level detection circuit calculates the fuel level from the fuel gauge which is present in all the vehicles. Connecting to internet requires additional money deducted by internet provider and the more use of battery. In order to be efficient, SMS function is included to start/stop internet connection. When the owner needs to track the vehicle, a SMS is sent to the device which then establishes the connection to the TCP server through GPRS. Data such as location speed and fuel level is sent to the server and stored in a database. The web page and android application designed for this purpose tracks the vehicle three dimensionally and show them in integrated map. Emergency call function is designed. The driver can call to the police, ambulance and the owner of the vehicle in emergency situations. The owner can dial a call to the vehicle at situations when the driver goes beyond the limits like high speed driving and wrong direction.

Keywords: Fleet Management, MSP430F5419A, Android application, Fuel level detector, GPS/GSM/GPRS.

Journal of Green Engineering, Vol. 4, 195–210.
doi: 10.13052/jge1904-4720.432

1 Introduction

Managing the fleet is difficult in these days. High fleet maintenance cost, security risks and raising fuel costs are the challenges faced by the owners of the fleet. Vehicle tracking system with fuel monitoring and emergency call function designed at low costs provides the solution for those problems. The challenges faced when designing this system includes size of the tracker device, cost of design, power constraint, server and software maintenance cost. The growing field of electronics provides us components very small in size, more efficient in working and is available at low costs. With these newest technologies a hardware device is designed.

2 Methodology

2.1 Hardware

SIM908 is a quad-band Global System for Mobile Communication (GSM) module which includes Global Positioning System (GPS) function for satellite navigation. Time and costs are saved because of its compact design which integrates General Packet Radio Service (GPRS) and GPS in a single SMT package. It allows tracking our assets seamlessly at any location and anytime with signal coverage. The input supply voltage range is 3.3 to 4.5 Volts. SIM908 is controlled via AT commands. The SIM908 is connected to micro-controller through Universal Asynchronous Receiver/Transmitter (UART). The integrated Transfer Control Protocol, TCP/IP stack makes it easy to establish TCP/IP connection. The data transfer supports two modes transparent mode and Non-transparent mode. In the transparent mode, the data received are sent serially to micro-controller automatically.

LP38501AT-ADJ is the linear voltage regulator. Flex Cap LDO's feature unique compensation that allows the use of any type of output capacitor with no limits on minimum or maximum ESR. This ultra-low dropout linear regulator responds very quickly to step changes in load, which makes them suitable for low voltage microprocessor applications. The input can be in the range of 3 volts to 5.5 volts. Output current is 3A. The output can be varied from 0.65 volts to 5 volts. The Voltage regulator has five pins and is surface mounted component. The adjust pin acts like feedback and helps the regulator to maintain the required output voltage effectively. The enable pin is used to start the voltage regulator.

TVS arrays are designed to protect sensitive electronics from damage or latch-up due to ESD and other voltage-induced transient events. SMF05C is the surface mounted IC which is very small is size and provides up to five connections which can be used in parallel to the circuit to protect from the electrostatic discharge.

The MSP430F5419A an ultra-low power microcontroller is used in this project. It has four UART, SPI and I2C interfaces which can be used to add more features in future. There are 100 in which 87 are general purpose input/output pins.

The port 1 and port 2 has internal pull up resistors connected which can be enabled when required. It has 128KB flash memory which can be used for storing the collected data as well as the code.

The target board MSP-TS430PZ5x100 is designed by Texas Instruments for the purpose of testing and developing MSP430 F5 series micro-controller and its applications. The board has a JTAG port which is connected to the MSP430 FET USB programmer. The USB programmer is used to dump the codes in the micro-controller and can act as a simulator. External crystal of 32 KHz is used in addition to the inbuilt 25 MHz crystal. Lower the frequency of the crystal higher the accuracy the will be[1].

ADS1110 is a 16 bit ADC which will give a more accurate output in digital format. This is continuously self – calibrating Analog to digital converter with differential inputs is very in size. The ADS1110 uses I2c serial interface and operates from a power supply ranging from 2.7V to 5.5V

The entire tracker device design can be made as small as the size of a human's palm when the components mentioned above are used. They are small and power efficient. A PCB with two to three layers will be required to make such a small device.

2.2 Software

The webpage is designed with the Hyper Text Markup Language (HTML) and Personal Home Page (PHP) language. MySQL is used as database software to store the details of the users. Each user can have up to five number of vehicle which can be tracked at the same time. The user has to login, in order to track the vehicle owned. The three dimensional vehicle information shows the vehicle's direction, altitude, latitude, longitude and speed.

The website "*www.dkarthik.net23.net*" is designed. The website has access to the web tracker application. The android application can be downloaded from the website. The website is hosted by *000webhost.com* and is 24x7 online. But, the tracker application link, "*www.myproject.ttl60.com*", provided under the *Products and Services/Online Tracker* of the created website is hosted by the home server in which the computer acts as the server. So, it works only when the computer used as the server is provided internet connection with IP address same as router's IP. If not port forwarding will be required to make server host the page online. The GPS Gate Server uses Internet Information Service (IIS) to host the webpage designed. IIS can only host one page at a time. Unlike Apache which can host more than one webpage at a time.

With the android application designed the user can track their vehicles at any time and in any place with their smart phones. The integrated Google maps will show the user the location of the vehicle marked in red color. The android application is created in ECLIPSE software which allows the program to access the GPS data from the tracker device to the phone and start tracking the vehicle.

Code Composer Studio V5 is used to write the micro-controller's code for the tracker, fuel level detection and emergency call. Cadsoft Eagle V6.3 is used to design the PCB.

2.3 Working Methodology

When the SMS is received, the GPS module starts collecting the NMEA data and sends it to the GSM. The NMEA data has latitude, longitude, speed, altitude, time date and direction of the vehicle. The collected data is sent to the server through a GPRS internet connection. A TCP/IP connection is established between the server and the Device. The data sent to the Gps gate server is stored in a database of that particular user which is identified by

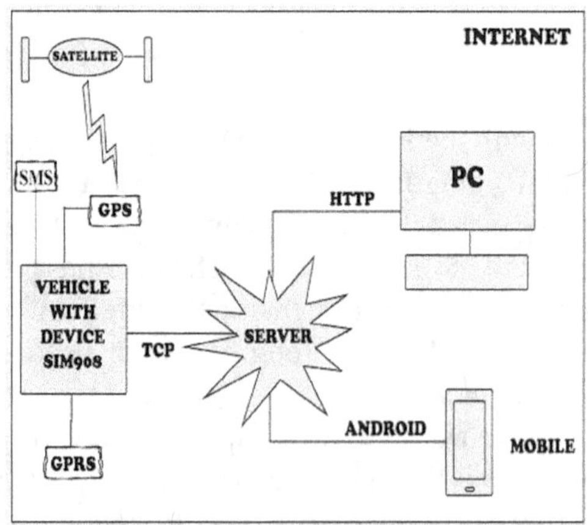

Figure 1 Working methodology

the IMEI number. User has to login to the website. The hosted website has embedded Google maps in vehicle can be tracked in real time. The connection can be closed with command given from website or through SMS. The micro-controller runs in the low power mode to reduce the power requirements by disabling interrupts and crystal oscillators.

3 System Structure

The six pin SIM card holder, net light indicator microphone and speaker are connected to SIM908 with the supporting circuits as suggested in the hardware design manual.

The net light indicator shows the network registration status of SIM card. The micro-controller is connected serially through UART to the SIM908 device. Both the device and modem supports Complementary Metal Oxide Semiconductor (CMOS) level hence no level shifters are required. The SIM908 is Data Communication Equipment (DCE) whereas the micro-controller is the Data Terminal Equipment (DTE). The Request to Send (RTS) and Clear to Send (CTS) line are data hardware flow control lines. When the micro-controller has to send the data the RTS is enabled and the modem acknowledges back with the enabled CTS signal. The Ring Indicator (RI) will be enabled for about 300ms when a call is received. Data Terminal Ready (DTR) is enabled has to be pulled down to show the DCE that DTE is ready to communicate.

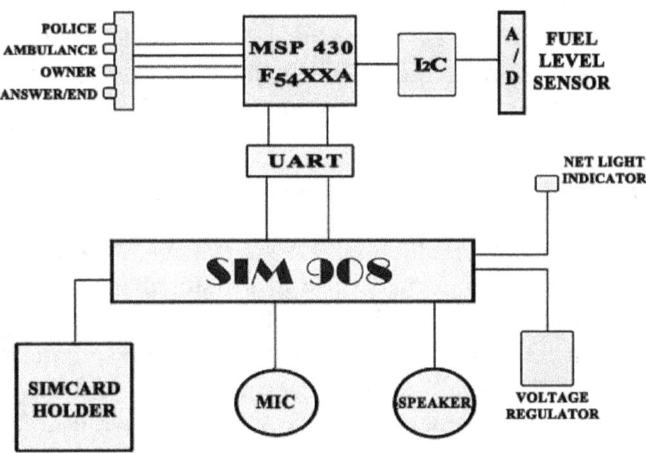

Figure 2 Basic structure of system

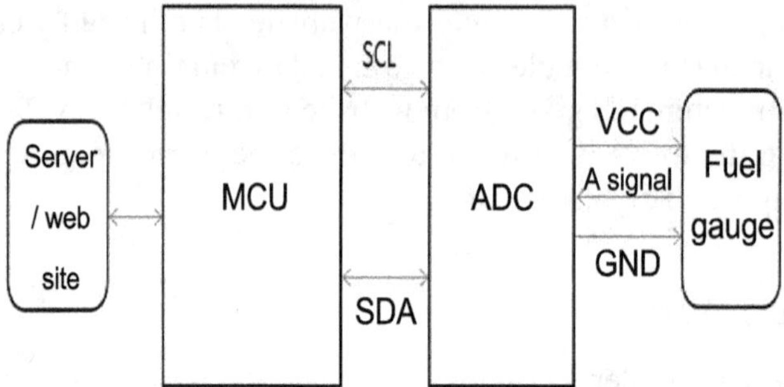

Figure 3 The overview of fuel level detector

3.1 Fuel Level Detection

The Analog to Digital (A-D) converter is connected to the micro-controller through Inter-Integrated Circuit (I2C) interface. The input to the converter comes from the fuel gauge kept inside the vehicle. Working of the A-D converter requires micro-controller programming. When the analog data is received by the A-D converter, the digital data is sent to the micro-controller and stored in a character array which is then converted to the fuel level in liters.

3.2 Emergency Call

Four button switches are connected to the second port of the micro-controller with the internal pull up resistors enabled. When the switches are pressed the corresponding calls are made to the respective people such as police, ambulance and owner of the vehicle. The switches can be set on the dash board of the vehicle, whereas the device is embedded inside the vehicle.

4 Microcontroller Programming Methods

The commands given to the SIM908 device starts with "AT+". Hence, the commands are called AT commands. These are stored inside the micro-controller and have to be sent to the SIM908 during the flow of the program upon meeting certain conditions. AT commands are stored in character array and sent serially to the SIM908 device. Each command sent the SIM908 acknowledged back with characters "OK". If some other characters are received then there is some error in the code or real difficulties caused in real time.

4.1 Emergency Call Algorithm

AT Commands are sent from micro-controller to the SIM908 device. Once an AT command is sent, next command is sent only when positive acknowledgement "OK" is received.

1. Initialize the variables, stop watch dog timer and enable interrupts.
2. Send the command "AT" to check the communication is successful. Modem replies back "OK"
3. Set the baud rate of 115200 by sending "AT+IPR=115200". After this command the program enters into the interrupt.
4. To answer the call the command "ATA" is sent when the button4 is pressed when the device is ringing. This same button is used for ending an ongoing call. "ATH" is sent to end the call when the button is pressed during the call period.
5. To dial a particular number the command "ATD<number>" is sent. Mobile number of the person to be called should be given instead of <number>. Three numbers should be called for three different buttons. Button1, dials to ambulance, Button2, dials to police, Button3, dials to owner of the vehicle.

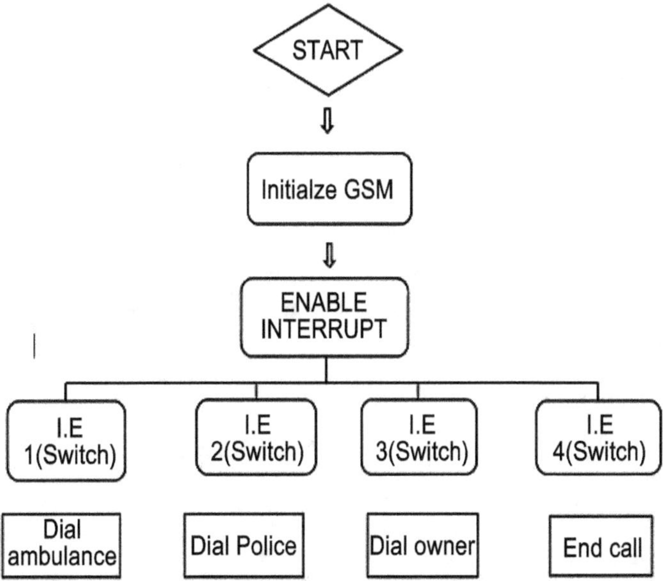

Figure 4 The working of emergency call function

4.2 GPS Data Collections Algorithm

Each command is sent only the positive acknowledgement “OK” is received for the previous command.

1. When SMS message containing “START” is received power on the GPS device with the command “AT+CGPSPWR=1”.
2. First time after the device is powered on, the device has to be reset in cold mode. It takes 40 seconds in good satellite coverage to fix the location three dimensionally. Hence, the command “AT+CGPSRST=1” is sent. Waits for 40 seconds after sending this command.
3. To check whether the location is fixed or not. “AT+CGPSSTATUS” command is in a time interval of 20 seconds until location fixed acknowledgement is received.

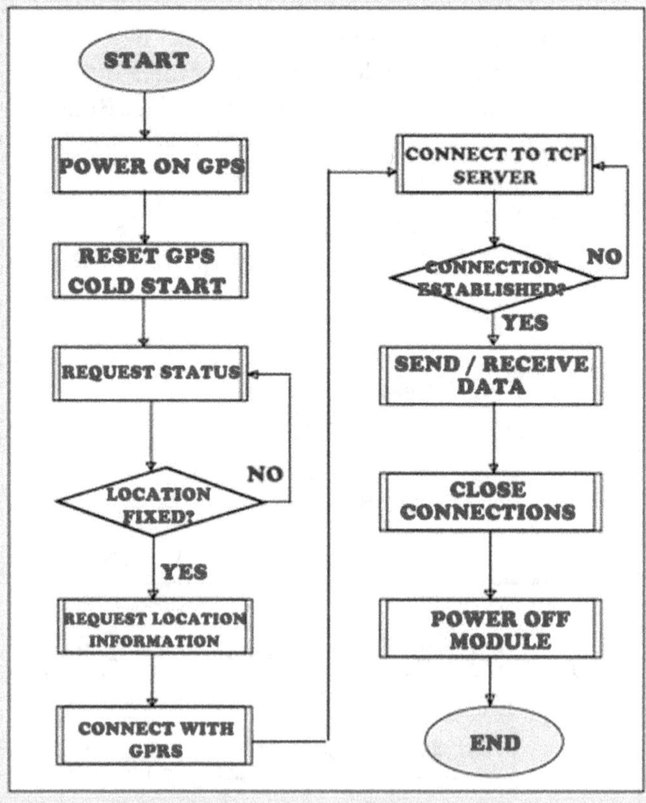

Figure 5 Flowchart of GPS and GPRS

4. Once the location is fixed, to get current GPS location the command "AT+CGPSINF=32" is sent. There are different formats in which the GPS NMEA data is received from the modem. The required format is chosen and the command for the particular format has to be sent to acquire the NMEA data.
5. The acquired data is not in the format the server needs. The format needed by the server is also not in the list of supported formats. Hence re-arranging, the acquired data is to be done and the check sum has to be added before the data is sent to the server.
6. By calculating the number of commas, the data can be stored in the different character array and can be merged together in the single character array. This is achieved by using number of for loops and conditional statements.

4.3 GPS Tracker One Server Protocol

4.3.1 Start tracking

This command is sent from the server to the tracker device. It has rules. The rules like at which time interval the device has to keep sending the location information[4]. The syntax of start tracking will be as follows

"$FRCMD,IMEI,_StartTracking,,Rule1=value1,Rule2=value2,...*XX"

4.3.2 Stop tracking

After the start command, the device keeps sending the location of the vehicle at certain time intervals. This command should stop the tracking. When this command is received, the micro-controller program should be written in a way to stop the Tracking and turn off the GPS and GPRS connections and enter low power mode[4]. Has syntax as follows

"$FRCMD,IMEI,_StopTracking*XX"

4.3.3 Send message

This command is sent from the micro-controller to the server. GPS data and fuel level is sent to server with this command[4]

Syntax is as follows

"$FRCMD,IMEI,_SendMessage,,DDMM.mmmm,N,DDMM.mmmm,E,AA.a,SSS.ss,HHH.h,DDMMYY,hhmmss.dd,valid,var1=value,var2=value...*XX"
Where "*XX" is the checksum.

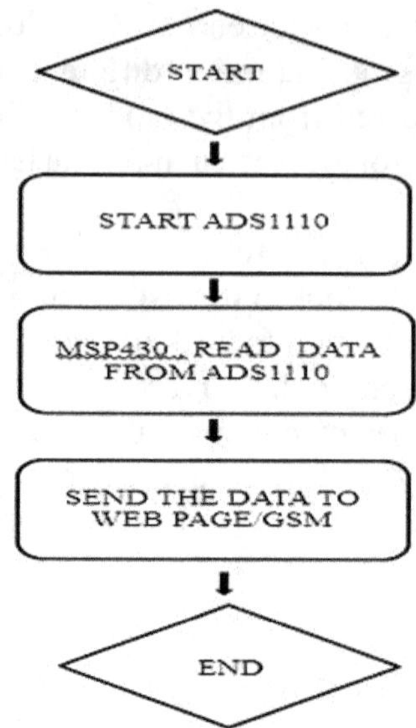

Figure 6 Flowchart of fuel level sensor mechanism

4.4 Fuel Level Algorithm

1. Initialize the variables in unsigned integer.
2. Initialize i2c clock, set baud rate, enable interrupt.
3. Generate an acknowledgement and Not Acknowledgement signal.
4. Start I2C bus.
5. Stop I2C bus.
6. Read data from ADS1110 Data register and convert it with the fuel level value in liters.
7. Store the fuel level in the character array "fuel []". This is sent later with the send message command to the server.

4.5 GPRS Algorithm

1. To set the GPRS active, the command "AT+CGATT=1" is sent.
2. Start the GPRS connection by the command "AT+CGDCONT=1,"IP"," <apn_name>""". <apn_name> is provided by the internet service provider.
3. To start task and set APN, User name, password the command "AT+CSTT="<apn_name>,"<username>", "<password>"""" is sent to SIM908.

4. To bring up wireless connection with GPRS the command "AT+CIICR" is sent.
5. To get local IP Address, the command "AT+CIFSR" is sent.
6. For knowing the current connection status the command "AT+ CIPSTATUS" is sent. If reply is "CONNECT OK" the GPRS connection is successful.
7. To add an IP head at the beginning of package received "AT+CIPHEAD" is sent.
8. To start TCP connection, the command "AT+CIPSTART="TCP", <IP address>, <port>" is sent.
9. AT+CIPSTART" starts the TCP connection to the server.
10. Send the data through the TCP connection the command "CIPSEND" is sent. When ">" is returned by the SIM908, the data to be sent to the server should be written within five seconds. After five seconds the connection to the server will be closed.
11. If some data is received from server, it'll be sent to the micro-controller automatically by SIM908 because it's connected in transparent mode. From the received data, the required commands and data to the server are sent back using "AT+CIPSEND".

4.6 Android Applications

The application for the android phones is developed with the help of eclipse software and the language used is Java language[5]. Google map is integrated into the application. This is done by receiving API key from Google and generating a MD5 key from ECLIPSE[6]. Tracking the fleet involves the following steps

1. Java script code in android application sends the Asynchronous Java script (AJAX) request to the PHP script on the server every minute or on demand by the user using an event.
2. The PHP script will then obtain the last record for that user/vehicle from the database table and sends the latitude and longitude to the java script code running on the client side.
3. The application has embedded maps which shows the location of the vehicle

The application developed can be downloaded from the website "www.dkarthik.net23.net". This is the designed webpage to track the vehicle.

4.7 Web Page Design

The webpage is developed using HTML, scripting language like PHP and database by MySql. The Google map is embedded into the web application to view the fleet. These maps can be integrated by using the Google's API key. The web application is hosted in a home server whereas the website containing the information is hosted by "000webhost.com".

4.8 Result

The result obtained matched our design goals. PCB design worked properly without having much Electro-Magnetic interference and Electro-Static Discharge. The designed PCB and the assembly of the entire hardware is shown in the Figure 7. The GPS and GSM antennas are connected to the SIM908. The target board containing micro-controller is connected to the SIM908 in PCB board serially.

Figure 8 shows the tracked vehicle as red marker in the maps inside application website. The red marker marks the correct location *Kannamangalam*, near Vellore, where the device was set. The fuel level detected is displayed on the webpage in liters.

Figure 7 Hardware setup

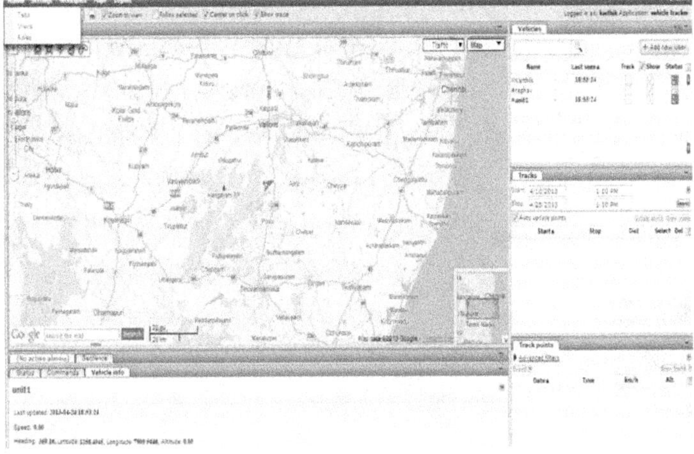

Figure 8 Vehicle tracker output

5 Conclusion

The emergency call function has been tested and implemented in real time. When the buttons are pressed the calls are dialed/received as per the project's motive. The GPRS connection and the SMS function are utilized successfully to start and stop the data usage and connect to the internet. The TCP connection can be established successfully with the server. Location and fuel level can be shown in the website as well as in the android application. Thus, the tracker device works in real-time. As the further development, more number of sensors can be included to know more about the vehicle's condition and more SMS functions can be added. So that when the connectivity is lost the client can still know status about the vehicle.

References

[1] Sachin S. Aher and Kolkate R. D, "Fuel Monitoring and Vehicle Tracking Using GPS, GSM and MSP430F149" published - *international Journal of Advances in Engineering & Technology,* July 2012.

[2] Texas instruments web site and forum [Online] Available: http://www.ti.com

[3] The official Microsoft IIS site and forum [Online] Available: http://www.iis.net

[4] *SIMCOM's manual SIM908 Hardware Design V1.00*, June 2011
[5] The Gpsgate website and forum. [Online] Available: http://www.gpsgate.com
[6] The android application developer team. [Online] Available: http://www.developer.android.com

Biographies

M. Shanmugasundaram obtained a BE in Electrical and Electronics Engineering from Manonmaniam Sundaranar University, Thirunelveli, in 1999, and an M.E in Embedded System Technologies from Anna University, Chennai, in 2006. Currently he is a senior assistant professor in VIT University, Vellore. His research interests include scheduling in real time system and fault-tolerant computing.

D. Karthikeyan obtained his BTech degree in Electronics and Communications Engineering from VIT University, Vellore, in 2013. Currently he is working as software Engineer. His area of interests include real-time systems and networks.

K. Arul Prasath obtained his BTech degree in Electronics and Communications Engineering from VIT University, Vellore, in 2013. Now he is doing Master degree in Communication Engineering in the same university. His research interests include microcontroller, Automobile and communication network.

R. Sri Raghav obtained his BTech degree in Electronics and Communications Engineering from VIT University, Vellore, in 2013. Currently he is doing his master in Business Administration. His area of interests include software development for various platform.

Reliable and Energy Efficient Topology Control Algorithm Based on Connected Dominating Set for Wireless Sensor Network

Manisha Bhende and Sanjeev Wagh

Research Center, Pad. Dr. Y. Patil Institute of Engineering an Technology, University of Pune, India
Email: manisha.bhende@gmail.com; sjwagh1@yahoo.co.in

Received 10 January 2015; Accepted 15 January 2015; Publication 19 March 2015

Abstract

Energy consumption in wireless sensor network is of paramount importance, which is demonstrated by a large number of algorithms, techniques and protocols that have been developed to save energy and to extend lifetime. Poly is the topology construction protocol. It is based on the idea of a polygon. Lifetime extension is one of the most critical research issues in the area of Wireless Sensor Network due to the severe resource limitation of sensor nodes. One of the key approaches for prolonging the sensor network's operable lifetime is to deploy an effective topology control protocol. We propose a Topology control algorithm for intelligent and reliable clustering, introducing energy harvesting nodes for maximizing the lifetime of network by supporting energy backup in the sensor field, proper localization of base station in the field to minimize the communication distance between cluster heads and the base station. Our simulation results demonstrate that IPoly performs consistently better in terms of energy efficiency message overhead, Energy overhead and reliability.

Keywords: Reliable, Connected domination set, Energy Efficient, Static and Dynamic, Topology Maintenance.

Journal of Green Engineering, Vol. 4, 211–234.
doi: 10.13052/jge1904-4720.433

1 Introduction

Due to advancement in technologies and reduction in cost of technologies and reduction in size, sensors are becoming involved in almost every field of life. WSNs can be widely used such as agriculture, Industry, Medicine, Horticulture and Military [11, 39]. In mission critical application; packet loss is not acceptable. Generally it is assumed that packet, [23] when nodes in WSN are connected to their neighbor, there is a possibility of packet loss, and therefore reliability should be achieved while improving energy efficiency. Topology construction and maintenance are two phases of topology control. Topological property is established in the construction phase.

Connectivity should be maintained in the construction phase. Second phase is the topology maintenance phase. In CDS based Topology control scheme, some nodes [1, 7] are part of virtual backbone. Non CDS node conserves energy by turning off radios. To achieve reliability and energy efficiency CDS size is an important parameter. For small CDS network traffic is handled by very few nodes, resulting into draining the battery. This is disadvantages of CDS. The advantage of this system is more nodes can go to sleep mode. "Saving energy compromises reliability". Poly is semi distributed graph theoretic topology control protocol for WSN. It finds the number of polygon present in the network, by modeling network as connected graph. To achieve energy efficiency, the protocol forms a CDS like polyphonic network, which in turn provide reliability in the case of random link failure. It adapts to topological changes in the remaining energy of nodes. The problem of maximizing the wireless sensor network lifetime is broadly categorized into direct approach and indirect approach [37, 38]. In indirect approach minimize energy consumption is managed through various intelligent algorithms, while the other approach directly supports external support to maximize network lifetime. Though the indirect approach can help extend the network lifetime, it does not focus on the problem of maximizing network lifetime. [23] Wireless sensor devices are cheap devices with fairly high failure rates. Further, in many applications, these devices have to be thrown into the area of interest from a helicopter, or similar vehicle. As a result, several nodes break or partially breaks affecting their normal functionality. Node reliability is also affected by crucial levels of available energy.

Many energy harvesting devices are proposed to supplement for the battery power of sensors to manage the power and control. The energy harvesting enabled networks mainly focus on the power management issue to estimate the amount of energy that can be harvested in the future [17–19], so as to optimize duty cycles and the scheduling of tasks [20–22] to maximize system performance, such as latency [23].

Clustering Mechanisms

The idea of clustering is to select a set of nodes in the network to construct an efficient topology. The selection of neighbors can be made on various criteria, namely, energy reserve, the density of the network or node identifier. Unlike in power adjustment or power mode approaches, the clustering approach constructs a topology with hierarchical structures that are scalable and simple to manage. The advantage of clustering is that a certain task can be restricted to a set of nodes called cluster heads and they can be assigned for collecting, processing and forwarding packets from non-cluster heads. This mechanism provides an efficient network organization. Other attractive features of the clustering approaches include the load balancing and data aggregation or data compression offered for prolonged network lifetime. In some clustering approaches, the selection of the cluster heads remains fixed. Hence, cluster heads typically experience faster energy depletion because they are heavily loaded with various tasks [4, 5, 12]. This problem is overcome by randomizing the selection of cluster heads to distribute loads fairly among nodes in the network.

2 Related Work

In the literature, there has been some work that protects previously existent topology control algorithms. Waltenegu Dargie et al [2010] proposed topology Control protocol [1, 23]. The developed protocol enables nodes to exhaust their energy fairly. The algorithm tries to preserve shortest path connecting itself to nearby nodes and the minimum-energy paths. In [2] concept of distributed topology control algorithm to conserve energy is introduced. In this paper localized distributed Topology control algorithm is presented. It calculates optimal transmission power to active network connectivity. It reduces node transmission power to cover nearest neighbor. A node uses only the locally available information to determine nodes. The majority of work has been done in fault tolerant topology control algorithm to minimize the total power consumption [16, 22]. It provides k-vertex connectivity between two vertices. Mihaela Cardei et al [4] propose a new architecture to achieve minimum energy consumption by using k-approximation, centralized greedy, distributed and localized algorithm. It provides reliable data gathering infrastructure from sensors to super node. In [13] summary of recent research results on Topology control techniques for extending the lifetime of battery powered wireless network is given. To increase the network lifetime, the design of efficient topology control of communication is very important.

EBC is based on SNA (Social network analysis) and measure the importance of each node in the network. QoS is achieved by evaluating relationships between entities of the network (i.e. edges) and identifying different roles among them (e.g. Brokers, outliers) to control information flow, message delivery, latency, and energy dissipation among them. This algorithm is applicable in homogeneous network and proposes a different line of research: Topology control in terms of QoS requirement. Given a set of nodes performing specific task, e.g. sink node in environmental sensor networks. The topology control algorithm is to select from the target network appropriate logical neighbors' of the former nodes, namely a subset of the physical neighbors' of former node that can be used to perform application specific procedure, without the need of involving the rest of physical neighbors' during execution of these procedures. QoS based topology control algorithm selects a suitable set of logical neighbors' such that input QoS requirements can be satisfied. EBC is bidirectional, weighted topology control algorithm. It is compared with GG, RNG and closeness centrally. In [17] authors proposed a self stabilizing algorithm for efficient topology control in Wireless sensor networks. It reduces the transmission power of each node so as to maintain network connectivity while saving maximum energy. The goal of the optimization is to minimize the average path length from source to destination to minimize the transmitted power. In [28] authors proposed novel topology control solution on the concept of betweenness centrally. This information allows us to achieve high quality of service. By studying the available literature we have found that still energy optimization can be done in topology control. Figure 1(a, b) and Figure 2(a, b) shows the Polygon formation for Dense and Sparse environment.

3 Material and Method

In the developed Topology Control Protocol some nodes are selected from the given set to create virtual backbone. Let V and D be be the set of nodes D $\in V$ Where all nodes in V are in D. The set of nodes is one hop neighbor to other nodes

$$d \in D\,(\forall v \in V \nabla \exists d \in V\colon (v, d) \in E)$$

Redundancy is defined as the expected number of functional spanning trees in a graph G. Every edge must be considered as a bridge in the spanning tree. Network Reliability is achieved with the help of spanning tree. There should be one spanning tree in the network to handle random link failure. An adjacency matrix of a graph G is denoted by A = $(Ai.j)\,n.n$

Then

$$A_{i.j} = \begin{cases} 1 & if\ vertices\ vi\ and\ vj\ are\ adjacent \\ 0 & otherwise \end{cases}$$

The degree of vertices is represented by diagonal matrix. If D denotes diagonal matrix of graph G then

$$di.j = \begin{cases} \deg(vi) & for\ i = j \\ 0 & i \neq j \end{cases}$$

The energy harvesting node, transmit power control and maintain topology of the network as well as the achievable throughput of the network. The minimum number of sinks required to keep the network connected is analyzed [25].

4 Proposed Algorithm Working

The resulting topology obtained by IPOLY provides a desired level of packet delivery and energy consumption is less than CDS. It has low message overhead. Among a set of nodes, poly protocol forms a closed path. It [2] provides reliable and energy efficient topology because it allows nodes to use an alternative in case of random link failure. Position or orientation information is not considered by this protocol. For energy saving dormant nodes are entered into sleep mode.

A. Description of Control Messages

Three types of messages are used by Poly at the time of the polygon formation process:

- Hello
- Create topology
- Finish discovery

Parent id of the sender is contained in hello message. To announce the end of topology discovery – finish discovery message uses a create topology message containing the IDS of the active node set is propagated

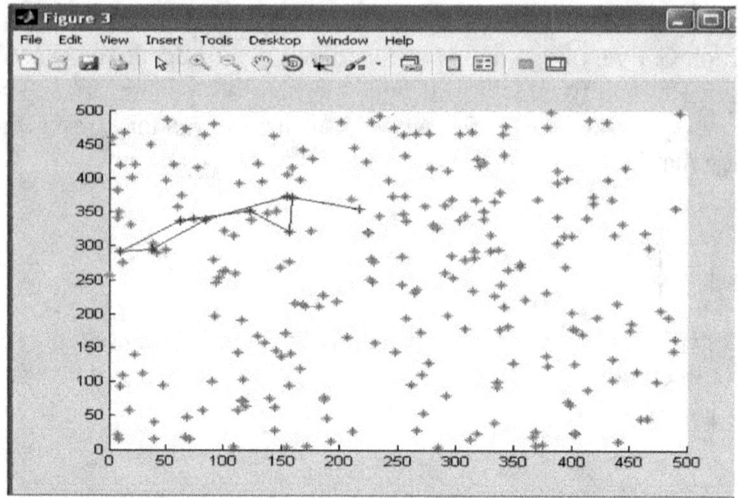

Figure 1 a) Polygon formation for dense network

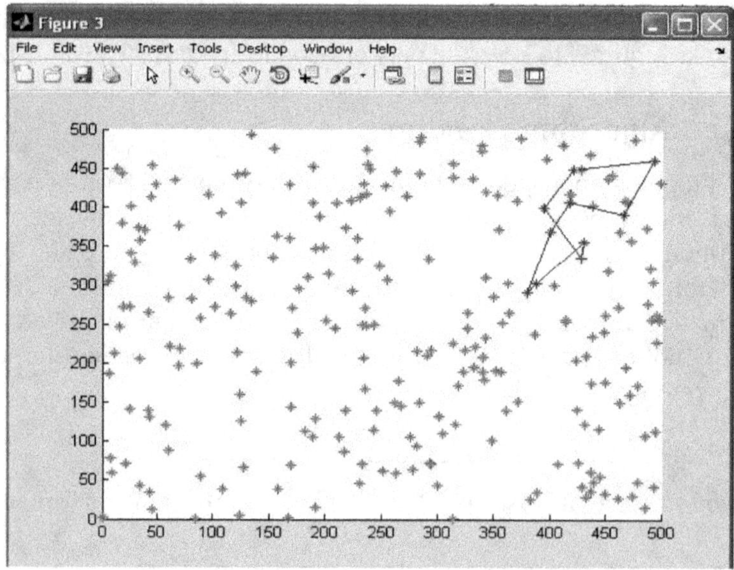

Figure 1 b) Polygon formation for dense network

in the network. Hassaan Khaliq Qureshi et el[23] proposed POLY [23]. We have consider the same assumption defined by the authors for implementation of protocols.

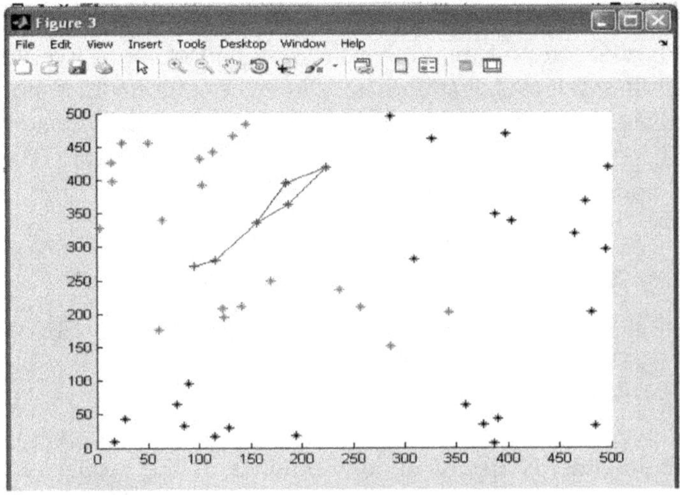

Figure 2 a) Polygon formation for sparse network

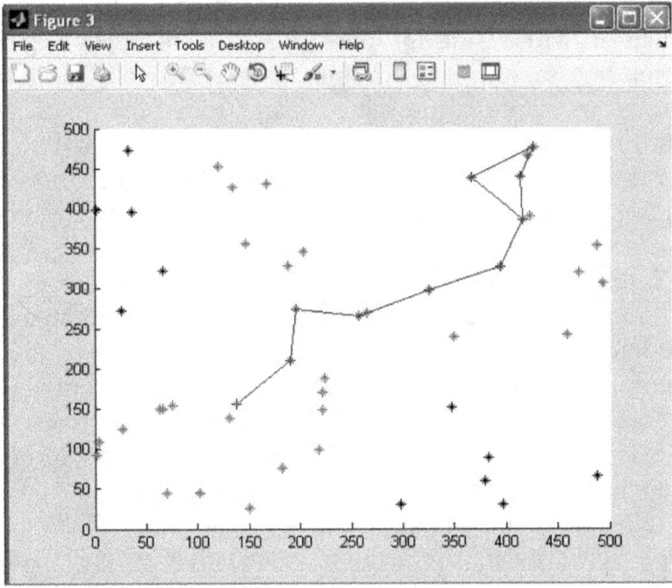

Figure 2 b) Polygon formation for sparse network

B. Topology Construction Protocol

The neighbor discovery process is initiated by sink node and CDS is created in this first phase of topology construction. In the second phase sink node received neighbor list. Discovery of polygons in the graph is done in the last phase. Polygon nodes are informed that they are part of the active node set [13]. Poly algorithm selected a random node as an initiator node. If more than one node initiates a process, the performance will be given to node having largest ID [38][39]. The hello of node A is received by B, F and H. These are the uncovered nodes.

C. Complexity Analysis of Poly and Proposed Protocol

Complexity of the CDS discovery process is same for A3, EECDS and CDS rule protocol but POLY have lower CDS discovery message complexity because it uses wireless broadcast for parent discovery. Figures 3–6 shows the comparison of CDS based protocols. In A3 children recognition messages contain ordered list of all the children of the sender. This list is used by children to set a timer to compete for an active node. When network is dense, this list increases with the increase in message size. And hence consume more energy. The more the children, the more length of message and it will result in more energy consumption per children recognition messages. Due to this reason, A3 uses a 100 byte size for children recognition message. Apart from other messages of size 25 bytes EECDS uses broadcast packet size of 25 bytes with 6 types of messages for topology construction, which does not exceed broadcast packet size, CDS rule K also uses 25 byte broadcast packet.

After CDS discovery A3, EECDS and CDS Rule protocols do not have any additional overhead, Poly introduces additional complexity. To reduce this additional complexity, Author discovers a subset of the cycle, they haven't considered all the cycles in the network. Therefore sink node processes a reduced subset of a message's path and few cycles. Author also utilized cycle merging smaller cycles are combined to form larger cycles. The additional complexity of poly protocol represents a trade-off between reliability and energy efficiency. The size of the polygon in the protocol is a critical parameter for evaluation of the algorithmic metrics is:

Message overhead- It is defined as total no of packets sent-received generated in the whole network during an experiment. Message overhead is directly proportional to energy consumption. Lower the message overhead, lower energy will be consumed. Every protocol designed in WSN is always trying to minimize this overhead.

Energy overhead- it is defined as the fraction of network energy expended during construction of topology. In case of topology maintenance this metric calculates overhead during reconstruction of topology under dynamic condition.

Residual energy- it is defined ratio of energy in the active set of nodes to the total network energy at the end of an experiment. Residual energy is a measure of network lifetime. As residual energy falls below a certain threshold value the probability of network partitioning increases.

Connectivity – connectivity refers to the number of nodes which are disconnected from the sink node after the activation of topology maintenance technique. This parameter measures the effectiveness of the topology construction protocol. If the connectivity value equals to zero, protocol is the best one. Higher value of connectivity shows that the protocol is unable to provide the backbone.

The message and energy overhead of EECDS, CDS-rule and A3 protocol compared with POLY. Among these three A3 has a low message and energy overhead due to its three way handshake protocol. Poly protocol has low energy overhead and greater message overhead than A3.A3 uses signal strength as selection metric for node selection in CDS. In grid topologies nodes are placed at equal distances which results in more energy overhead [11]. For a selection of node in proportion to the size of the network broadcast mechanism is used by poly. It results in better residual energy as compared to other protocol.

An increase in the node degree leads to an increase in the number of messages exchanged. Poly has been providing better residual energy because.

1) The active node set is proportional to network size.
2) Rebroadcast mechanism is used by poly; it consumes battery of node at an equal rate.
3) The active node set is proportional to network size.
4) Rebroadcast mechanism is used by poly; it consumes battery of node at an equal rate.

To reduce this additional complexity, discovers a subset of the cycle, haven't considered all the cycles in the network. Therefore sink node processes a reduced subset of a message's path and few cycles. Cycle merging has been done: smaller cycles are combined to form larger cycles. The additional complexity of poly protocol represents a tradeoff between reliability and energy efficiency. Figures 7–10 shows the Comparison of Static and Dynamic Implementation using energy harvesters. We compared results for message overhead, Energy overhead, connectivity and residual energy. The size of the polygon in the protocol is a critical parameter for evaluation of the algorithmic metrics:

In grid topologies nodes are placed at equal distances which results in more energy overhead [11]. For a selection of node in proportion to the size of the network broadcast mechanism is used by poly. It results in better residual energy as compared to other protocol. An increase in the node degree leads to an increase in the number of messages exchanged [40]. Poly has been providing better residual energy because.

5 Simulation Setup and Result

Poly is implemented in MATLAB environment. In the experiment we consider that sensor nodes are deployed in 600m*600m area randomly. 50 to 250 nodes are used to perform different network topology. We have considered energy based topology maintenance technique. Data packet size of 25 bytes and ideal Medium Access Control layer is used. There is no packet loss due to channel contention/collisions.

Following Matlab code shows the placement of energy harvesting nodes

```
function [node]=EHNode_Deployment(AREA,N,RR)

global DEPLOYED_NODES

NODE.ID=0;
NODE.CURRENT_ROLE='Energy Harvester';
NODE.DISTANCE=0;
NODE.POSITION=0;
NODE.RESERVED_ENERGY=2;
NODE.NBHRS=0;

%% Create random positions of the nodes
a=0;
x_cord = a + (AREA.X-a).*rand(N,1)-eps(RR/sqrt(5));
y_cord = a + (AREA.Y-a).*rand(N,1)-eps(RR/sqrt(5));
vertices=[x_cord y_cord];
clear a b x_cord y_cord

%% Use above Structure defination to create 'N' number of nodes
```

```
node=NODE;
node(N).ID=0;
for i=1:N
    node(i).ID=i;
    node(i).POSITION= vertices(i,:);
    node(i).CURRENT_ROLE='Energy Harvester';
    node(i).RESERVED_ENERGY=1;
end

%%SINK AT POSITION (Xmax,Ymax)
% node(N+1).POSITION= [AREA.X+20 AREA.Y+20];
%% DISCOVER NEIGHBOURS

Distance=[];
all_Distance=[];
for i=1:N
    a=node(i);
    for j=1:length(DEPLOYED_NODES)

d_ij=cal_dist(vertices(i,:),DEPLOYED_NODES(j).POSITION); %Cal Dist
b/w nodes
        if d_ij<=RR           %Whether the node is within range?
            a.NBHRS(j)=j;          %Add the id of node in neighbor list
                                                %    a.N_L(j)=1;
                                                %    Energy=[Energy e_ij];
        end
```

In static topology maintenance technique, performance is dependent on efficient topology construction protocol. We have considered the results of the dynamic topology maintenance technique based on energy threshold. Polygon size depends on the network size. 10 to 50 nodes are used to construct polygon.

Table 1 No of nodes and performance metric values

Sr No	No of Nodes	Msg Overhead	Residual Energy	Energy Overhead	Connectivity
1	50	1187.2	0.416	0.00534	18
2	100	4367	0.1512	0.00402	82
3	150	10266.2	0.08474	0.00274	136.2
4	200	17345.2	0.05912	0.0022	187.8
5	250	24234	0.04484	0.0018	240.4

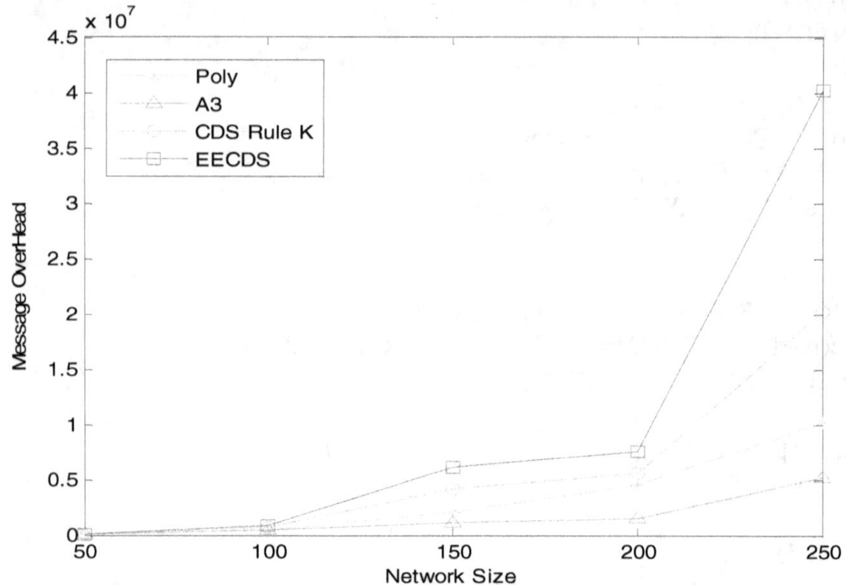

Figure 3 Comparison of message overhead

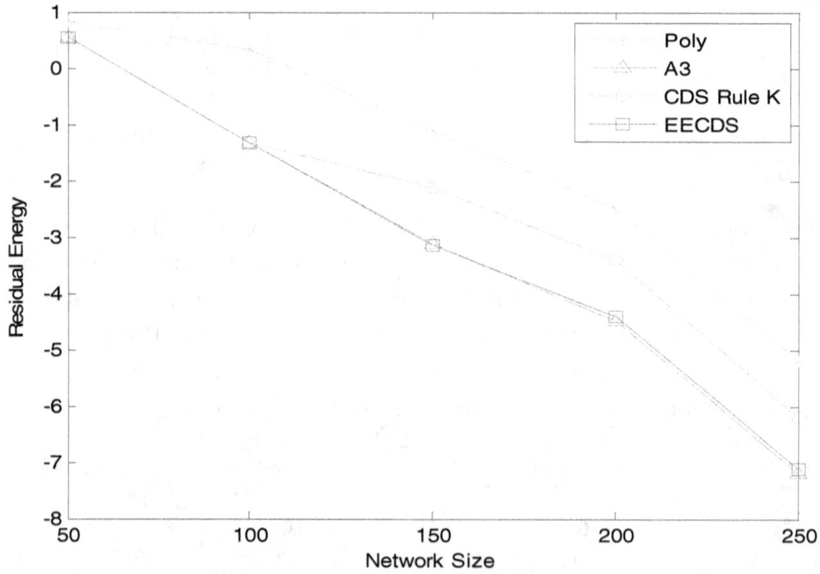

Figure 4 Comparison of residual energy

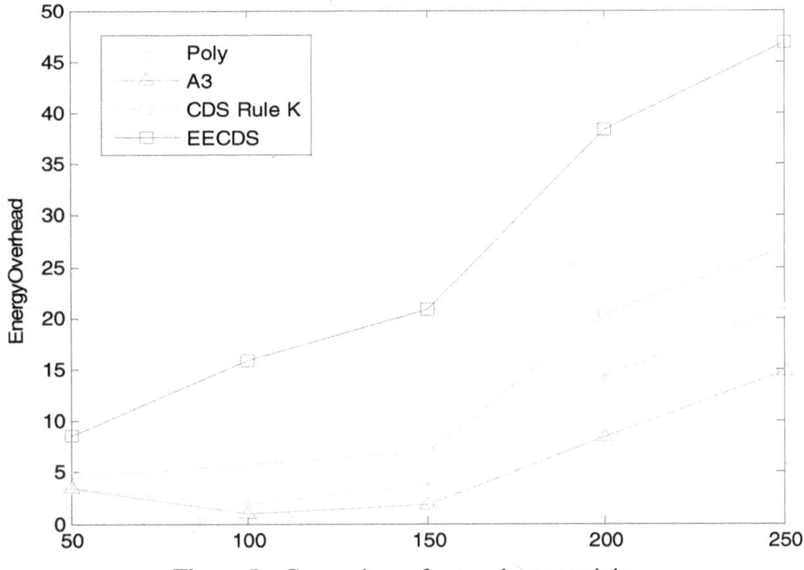

Figure 5 Comparison of network connectivity

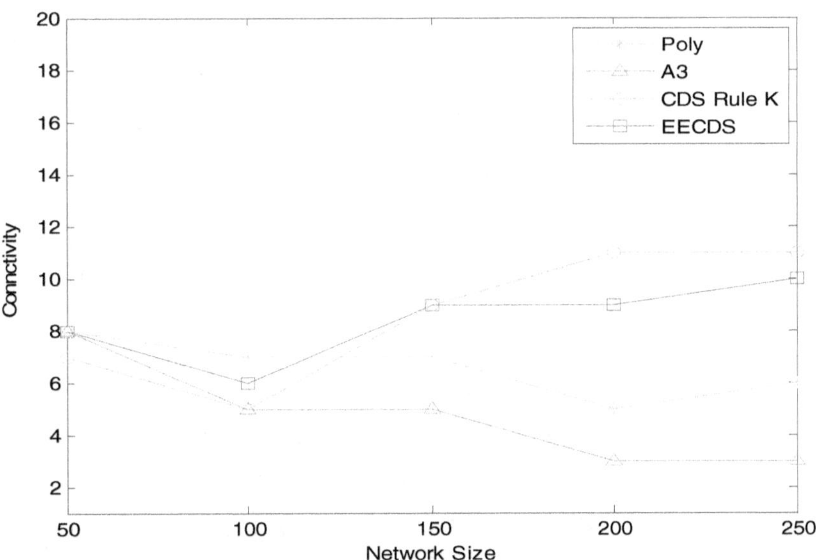

Figure 6 Comparison of connectivity

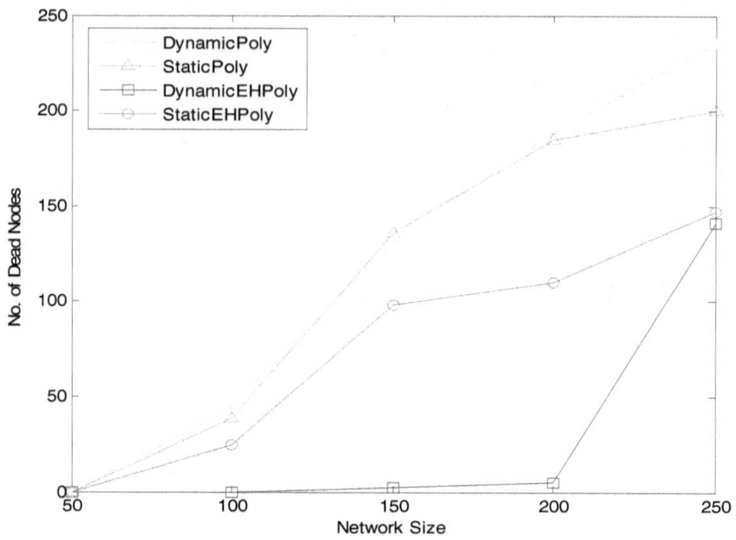

Figure 7 Comparison of network lifetime for static and dynamic topology control with energy harvester

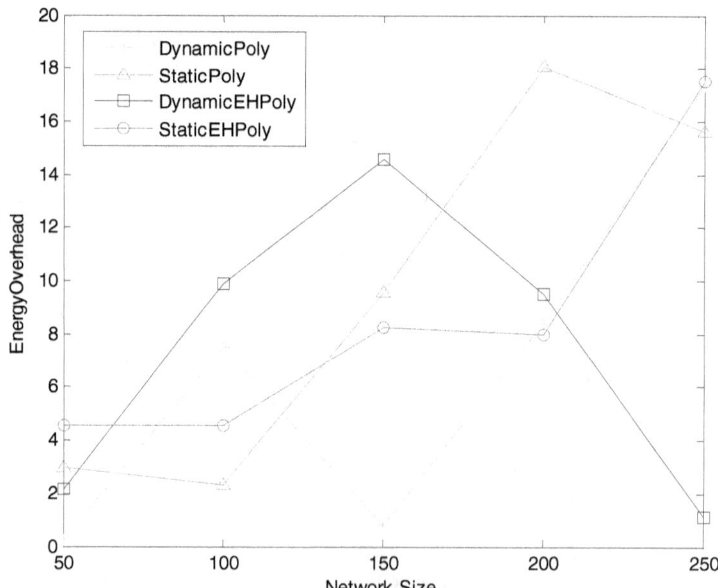

Figure 8 Comparison of energy overhead for static and dynamic topology control with energy harvester

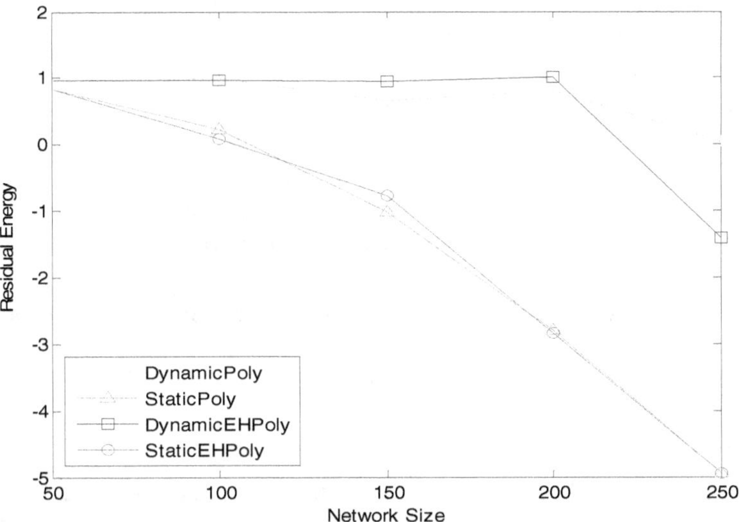

Figure 9 Comparison of residual energy for static and dynamic topology control with energy harvester

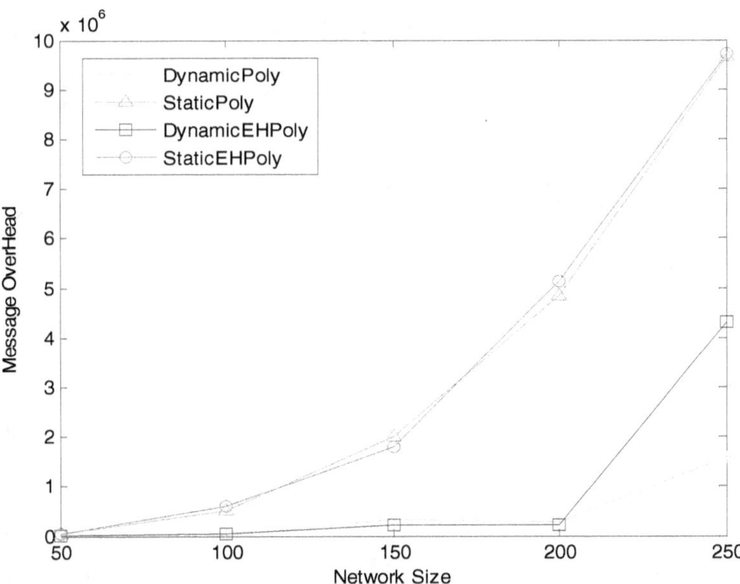

Figure 10 Comparison of message overhead for static and dynamic topology control with energy harvester

Controlled indoor deployments are consider to evaluate all five protocols. two ideal grid environment observed are:

The grid H-V.

The grid H-V-D topology.

In grid H-V nodes can communicate horizontally and vertically. In H-V-D nodes can be communicated horizontally, vertically and diagonally. Figures 11–14 shows the final comparison of proposed and the existion CDS based protocols. Result shows that Proposed protocol performs well in terms of message overhead and energy overhead and it is providing strong connectivity between the backbone nodes. So it is providing higher reliability with energy efficiency.

6 Conclusion

We computed the reliability for CDS based Poly protocol and compared with static and dynamic deployment of node deployment. By considering EH as a backup, the lifetime of wireless sensor network field can be maximize the lifetime of sensor fields with clustering and base station placement. Existing

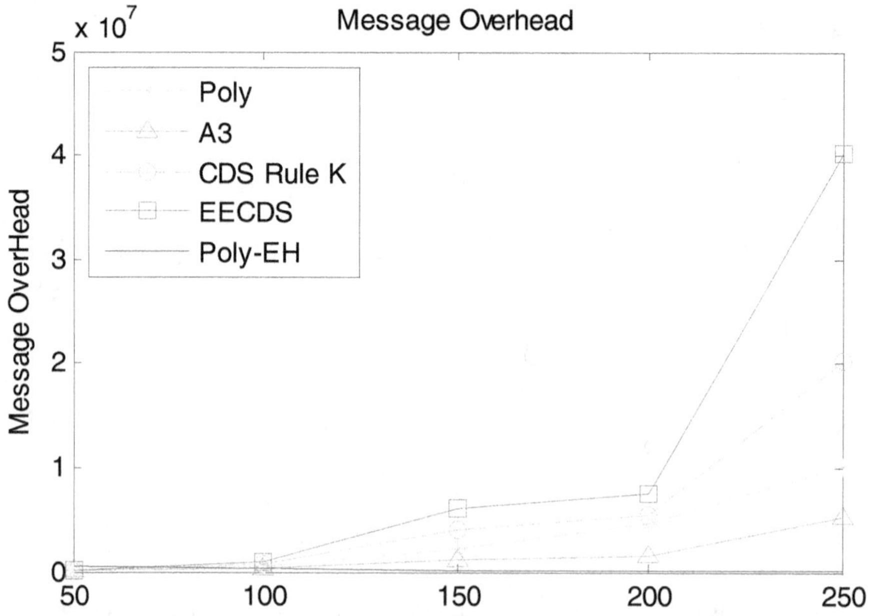

Figure 11 Comparison of poly and improved poly for message overhead

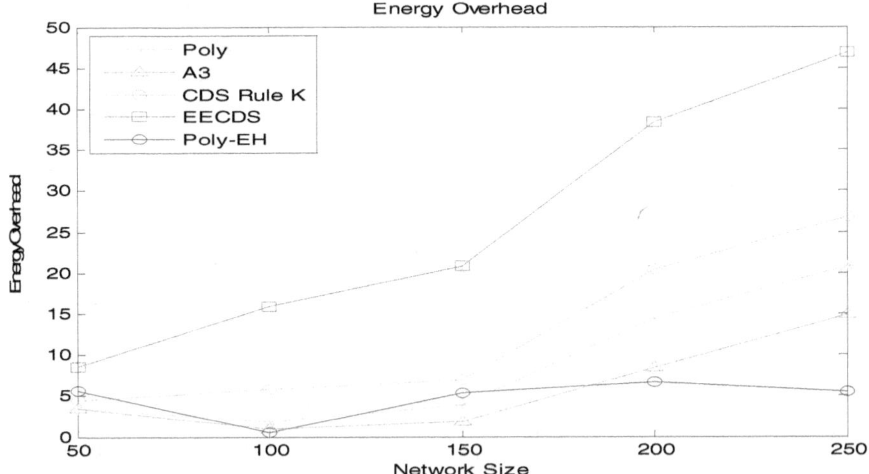

Figure 12 Comparison of poly and improved poly for message overhead

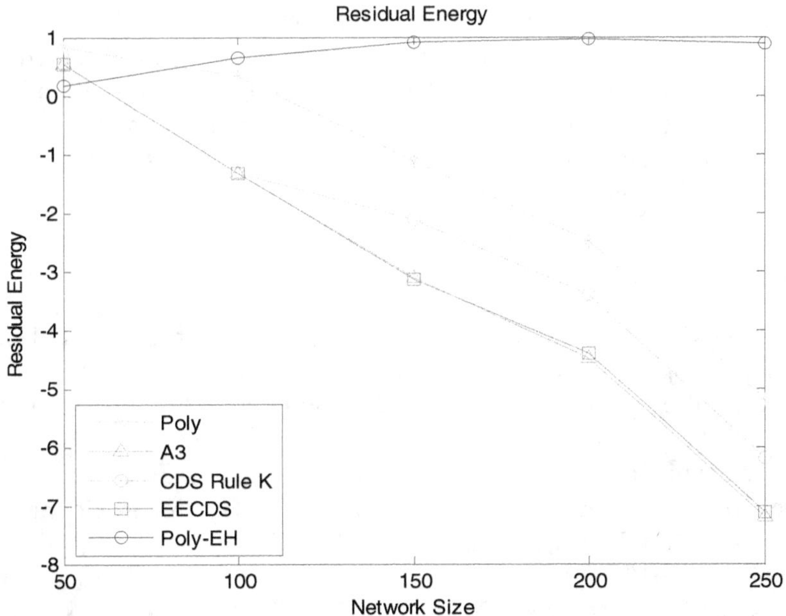

Figure 13 Comparison of poly and improved poly for residual energy

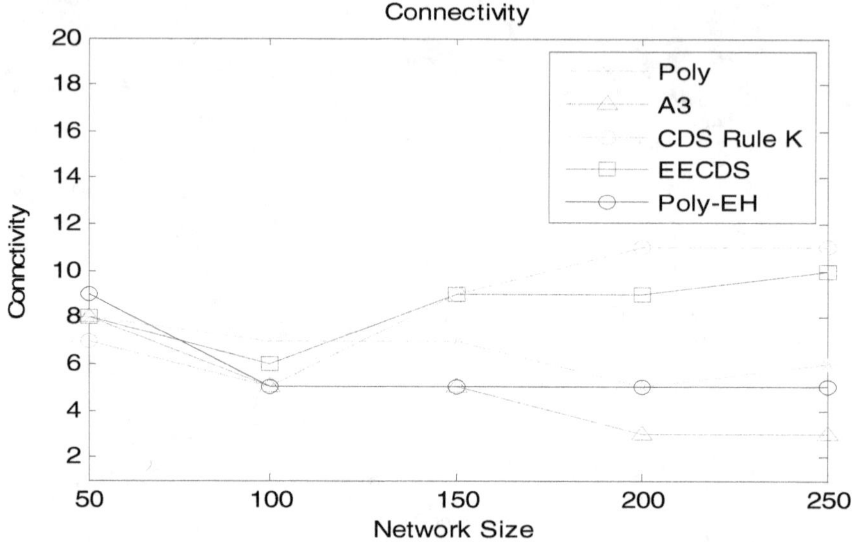

Figure 14 Comparison of poly and improved poly for connectivity

CDS-based protocols have considerably lower network reliability because each edge (link) in these topologies serves as a bridge edge, and therefore does not provide any redundancy in the network. Improved Algorithm has fewer message overhead. It has less energy consumption compare to other available CDS based technique. Developed algorithm performs well in static as well as dynamic environment.

References

[1] Waltenegus Dargiea, Rami Mochaourabb, Alexander Schill a, Lin-Guanc, "A topology control protocol based on eligibility and efficiency metrics," The Journal of Systems and Software, PP 1–10, August 2010.

[2] R. S. Komali, R. W. Thomas, L. A. DaSilva, and A. B. MacKenzie, "The Price of Ignorance: Distributed Topology Control in Cognitive Networks," IEEE Transactions On Wireless Communications, Vol. 9, No. 4, April 2010.

[3] Chi-Tsun Cheng, *Member,* Chi K. Tse, Francis, C. M. Lau "A Clustering Algorithm for Wireless Sensor Networks Based on Social Insect Colonies," IEEE Sensors Journal, Vol. 11, No. 3, March 2011.

[4] Mihaela Cardei, Shuhui Yang, Jie Wu, "Algorithms for Fault-Tolerant Topology in Heterogeneous Wireless Sensor Networks," IEEE TRANSACTIONS ON PARALLEL AND DISTRIBUTED SYSTEMS, PP 545–558 VOL. 19, NO. 4, APRIL 2008

[5] Tahiry Razafindralambo, Member, IEEE, and David Simplot-Ryl, Member, IEEE "Connectivity Preservation and Coverage Schemes for Wireless Sensor Networks," IEEE Transactions On Automatic Control, Vol. 56, No. 10, October 2011.

[6] Tao Shu and Marwan Krunz "Coverage-Time Optimization for Clustered Wireless Sensor Networks: A Power-Balancing Approach," IEEE/ACM Transactions On Networking, PP 202–215 Vol. 18, No. 1, February 2010.

[7] Ioannis Ch. Paschalidis, *Senior Member, IEEE*, and Binbin Li, *Student Member, IEEE* "Energy Optimized Topologies for Distributed Averaging in Wireless Sensor Networks" IEEE Transactions On Automatic Control, PP 2290–2304 Vol. 56, No. 10, October 2011.

[8] Pham Tran Anh Quang and Dong-Sung Kim, *Member, IEEE* "Enhancing Real-Time Delivery of Gradient Routing for Industrial Wireless Sensor Networks" IEEE Transactions On Industrial Informatics, Vol. 8, No. 1, February 2012.

[9] Yunhuai Liu, Qian Zhang, and Lionel M. Ni, "Opportunity Based Topology Control in wireless sensor network" IEEE Transactions On Parallel And Distributed Systems, Vol. 21, No. 3, March 2010.

[10] Antonio-Javier Garcia-Sanchez, Felipe Garcia-Sanchez, Joan Garcia-Haro, "Wireless sensor network deployment for integrating video-surveillance and data-monitoring in precision agriculture over distributed crops", Computers and Electronics in Agriculture 75 (2011) 288–303.

[11] Tapiwa M. Chiwewe, *Student Member, IEEE*, and Gerhard P. Hancke, *Senior Member, IEEE* "A Distributed Topology Control Technique for Low Interference and Energy Efficiency in Wireless Sensor Networks" IEEE Transactions On Industrial Informatics, Vol. 8, No. 1, February 2012

[12] Hiroki Nishiyama, *Member, IEEE,* Thuan Ngo, Nirwan Ansari, *Fellow, IEEE,* and Nei Kato, *Senior Member, IEEE* "On Minimizing the Impact of Mobility on Topology Control in Mobile Ad Hoc Networks" IEEE Transactions On Wireless Communications, PP 1158–1166 Vol. 11, No. 3, March 2012.

[13] Azrina Abd Aziz, Y. Ahmet Şekercioğlu, Paul Fitzpatrick, and Milosh Ivanovich "A Survey on Distributed Topology Control Techniques for Extending the Lifetime of Battery Powered Wireless Sensor Networks," IEEE Communications Surveys & Tutorials, PP 121–141, Vol 15, No.1, First quarter 2013.

[14] Yunhuai liu, lionel ni, and chuanping hu, "A Generalized Probabilistic Topology Control for Wireless Sensor Networks," IEEE JOURNAL ON SELECTED AREAS IN COMMUNICATIONS, PP 1780–1788 VOL. 30, NO. 9, OCTOBER 2012

[15] Ningxu Cai, Mohammad Golami, Robert Bennan, "Application Oriented Intellegent Middleware for Distributed Sensing and Control," IEEE TRANSACTIONS ON SYSTEMS AND CYBERNATICS, PP 947–955 VOL. 5, NO.11, NOVEMBER 2012.

[16] Sajjad Rizvi, Hassan Khaliq Querishi, "A1: An Energy Efficient Topology Control Algorithm for Connected Area Coverage in Wireless Sensor Networks," Journal of network and Computer Application, PP 597–605, NO. 53, NOVEMBER 2011.

[17] Jalel ben Othman, Karim Bessaoud, Alain Bui, Laurence Pilard, "Self Stabilizing Algorithm for Efficient Topology Control in Wireless Sensor Networks," Journal of Computational science, PP 1–10, NO. 117, JANUARY 2012.

[18] Hui Wang, Nazim Agoulmine, Maode Ma,Yanliang Jin, "Network Lifetime Optimization in Wireless Sensor Networks", IEEE JOURNAL ON SELECTED AREAS IN COMMUNICATIONS, PP 1127–1136 VOL. 28, NO. 7, SEPTEMBER 2010

[19] Yang Xiao, Miao Peng, John Gibson, Geoffrey G. Xie, Ding-Zhu Du, Athanasios V. Vasilakos, "Tight Performance Bounds of Multihop Fair Access for MAC Protocols in Wireless Sensor Networks and Underwater Sensor Networks," IEEE TRANSACTIONS ON MOBILE COMPUTING, PP 1538–1553 VOL. 11, NO. 10, OCTOBER 2012

[20] Stefania Sardellitti, Sargio Barbarossa, Ananthram Swami "Optimal Topology Control and Power Allocation for Minimum Energy Consumption in Consensus Networks," IEEE TRANSACTIONS ON SIGNAL PROCESSING, PP 383–399 VOL. 60, NO. 1, JANUARY 2012.

[21] Masoumeh Haghpanahi, Mehdi Kalantari, Mark Shayman, "Topology control in large-scale wireless sensor networks: Between information source and sink" Journal of, Ad Hoc Networks, article in press NOVEMBER 2012.

[22] Chuan Zhu, Chunlin Zheng, Lei Shu, Guangjie Han, "A survey on coverage and connectivity issues in wireless sensor networks," The Journal of Network and Computer Application, 35, PP 619–632, December 2011.

[23] Hassaan Khaliq Qureshi, Sajjad Rizvi, Muhammad Saleem, Syed Ali Khay, m Veselin Rakocevic, Muttukrishnan Rajarajan, "Poly: A reliable and energy efficient topology control protocol for wireless sensor networks," The Journal of Computer Communications, 34, PP 1235–1242, January 2011.

[24] Luo Xiaoyuan a, Yan Yanlin a, Li Shaobao b, Guan Xinping, "Topology control based on optimally rigid graph in wireless sensor networks" The Journal of Computer Networks, 34, PP 1235–1242, DECEMBER 2012.

[25] Linfeng Liu, Ruchuan Wang, FuXiao "Topology control algorithm for underwater wireless sensor networks using GPS-free mobile sensor nodes" The Journal of Computer Network and Application, 35, PP 1953–1963, August 2012.

[26] Halit Üster, Hui Lin "Integrated topology control and routing in wireless sensor networks for prolonged network lifetime" The Journal of Ad Hoc Network, 35, PP 835–851, September 2010.

[27] De-gan Zhang, Ya-nan Zhu, Chen-peng Zhao, Wen-bo Dai "A new constructing approach for a weighted topology of wireless sensor networks based on local-world theory for the Internet of Things (IOT)" Computers and Mathematics with Applications, 64, PP 1044–1055, Dec 2012.

[28] Alfredo Cuzzocrea, Alexis Papadimitriou, Dimitrios Katsaros, Yannis Manolopoulos "Edge betweenness centrality: A novel algorithm for QoS-based topology control over wireless sensor networks," Journal of Network and Computer Applications, 64, PP 1210–1217, July 2011.

[29] J. Guadalupe Olascuaga-Cabrera, Ernesto López-Mellado, Andres Mendez-Vazquez, and Félix Francisco Ramos-Corchado "A Self-Organization Algorithm for Robust Networking of Wireless Devices," IEEE Sensors Journal, Vol. 11, No. 3, March 2011.

[30] P. J. Wan, K. M. Alzoubi, O. Frieder, Distributed construction of connected dominating sets in wireless ad hoc networks, IEEE INFOCOM (2002).

[31] H. K. Qureshi, S. Rizvi, M. Saleem, S. A. Khayam, V. Rakocevic, and M. Rajarajan, An Energy Efficient Clique-based CDS Discovery Protocol

for Wireless Sensor Networks, in: proceedings of 44th Annual conference on Information Sciences and Systems(CISS), March, 2010.
[32] M. Gerla, J. T. Tsai, Multicluster mobile multimedia radio network, ACM Wireless Networks 1 (3) (1995) 255–265.
[33] P. M. Wightman and M. A. Labrador, A3: A Topology Construction Algorithm for Wireless Sensor Network, in: Proceedings IEEE Globecom, 2008.
[34] Z. Yuanyuan, X. Jia, and H. Yanxiang, Energy efficient distributed connected dominating sets construction in wireless sensor networks, in: Proceedings of the ACM International Conference on Communications and Mobile Computing, pp. 797–802, 2006.
[35] J. Wu, M. Cardei, F. Dai, S. Yang, Extended dominating set and its applications in ad hoc networks using cooperative communication, IEEE Transactions on Parallel and Distributed Systems 17 (8) (2006) 851–864.
[36] F. Wang, M. T. Thai, D. Z. Du, On the construction of 2-connected virtual backbone in wireless network, IEEE Transactions on Wireless Communications 8 (3) (2009) 1230–1237.
[37] Y. Wu, F. Wang, M. T. Thai, Y. Li, Constructing k-connected m-dominating sets in wireless sensor networks, in: Military Communications Conference, Orlando, FL, October 29–31, 2007.
[38] Sanjeev Wagh, Ramajee Prasad, 2013, “Heuristic Clustering for wireless sensor networks using genetic approach”, International Journal of Wireless and Mobile Networking (IJWAMN) Vol. 1, No. 1(November 2013).
[39] Sanjeev Wagh, Ramajee Prasad, 2013, “Power backup density based clustering algorithm for maximizing the lifetime of wireless sensor network”, IEEE Xplore.
[40] Wagh S., Prasad R: “Energy Optimization in wireless sensor network through natural science computing: A Survey”, Journal of Green Engineering, Vol. 3, No. 4, 01.07.2013, p. 383–402., DOI: 10.13052/jge1904–4720.342.

Biographies

M. Bhende received her ME from University of Pune and bachelors degree from government college of engineering, Amravati, India. Currently She is pursuing PhD from University of Pune. Her research interest include Wireless Sensor Network, Network Security, Cloud Computing and Operating system. She has published more than 20 papers in International, National conferences and Journals. She is working as reviewer for various International conferences and Journals.

Prof. Dr. S. Wagh received his Bachelors & PhD in Computer Science & Engineering from SRT Marathwada University, Nanded and Masters degree from University of Pune, India. He was guest researcher at Center for TeleInfrastruktur (CTIF) at Aalborg University (AAU), Denmark during March 2013 to May 2014. His research interest includes Computer Networks, Network Algorithmics, wireless sensor networks, IoT etc. He is currently involved in the research work of intelligent wireless communication, evaluation and energy optimization in wireless sensor networks. Sanjeev Wagh is a member of IEEE, Fellow IE, LMISTE & Fellow IETE. He is active technical committee member for the various top-quality conferences and journals in wireless networking.

Effect of Thermal Hydrolysis on Anaerobic Digester Performance

Leyla Amiri, Mohammad Ali Abdoli and Naser Mehrdadi

Department of Environmental Engineering, University of Tehran,
#15 Ghods St, Enghlab Ave, Tehran, Iran
Corresponding Author: Leyla Amiri <Leila.amiri63@gmail.com>

Received 6 May 2014; Accepted 2 November 2014;
Publication 19 March 2015

Abstract

Anaerobic digestion (AD) has been recently introduced as a promising technology for producing bioenergy around the world. Due to its especial characteristics, solid waste could be considered as a proper feedstock in the digester. Retention time (RT) and rate of biogas production are two major parameters affecting the efficiency of the process. The AD procedure of biogas production consists of hydrolysis, acidogenesis, acetogenesis, and methanogenesis such that feeding materials would be processed inside the digester. The experiments have been carried out using two bioreactors with volume of 5 liters in mesopihilic contitions at 35 $\pm$ 1°C. The purpose of this study is to accelerate the hydrolysis step. Pretreatment of the waste is accomplished using thermal hydrolysis in conditions of 155°C and 5 atm. The results of this research indicated 33.33% decrease in RT and 154.87% increase in amount of biogas production using pretreatment process in comparison without considering it.

Keywords: Anaerobic Digester, Organic solid waste, Thermal Hydrolysis, Biogas.

Journal of Green Engineering, Vol. 4, 235–244.
doi: 10.13052/jge1904-4720.434

1 Introduction

The amount of solid waste generated in Tehran, as a capital city of Iran, has been increasing steadily over the last decade. In 2012, the average amount of waste produced was 2.7 million tons/year [1]. Therefore, innovative solutions require diminishing its impacts on environmental quality and public health. Anaerobic processes have been widely used in waste treatment for over a century. Nowadays, it is also used for energy recovery from organic wastes such as garbage by methane production using one- or two-phase fermentation systems [2, 3]. Also anaerobic digestion (AD) has been used worldwide for over 100 years to stabilize wastewater solids [4]. Global concerns over energy security and greenhouse gases have given rise to the search for alternative energy source [5].

Developing biomass-to-biofuel technology could be significantly beneficial for enhancing energy security, reducing greenhouse gas emissions, and utilizing the renewable resources [6].

AD is extensively used as an alternative energy source from various biodegradable wastes. The AD produces biogas includes a methane and CO_2 which is suitable for energy production. The byproduct, from the AD process can be used as fertilizer for agriculture. Methane, a valuable source of energy, is the primary biogas produced by anaerobic processed [7]. AD process is a well established process for treating many types of organic wastes, both solid and liquid [8–10].

In order to the best manage of anaerobic digesters should put them under constant state conditions. For example, various parameters affect on its performance. These factors include pH, temperature, retention time (RT), loading rates and characteristic of substance.

AD of municipal solid waste were widely used in the world [11–13], but fewer research was done on enhancement of AD by using thermal hydrolysis.

There are numerous methods for pretreatment of solid particles in AD such as mechanical crumbling methods, chemical treatment, ultrasonic technique; thermal pretreatment, enzymatic pretreatment, electrical and freeze/thaw methods [14–17]. Thermal hydrolysis is a well proven method to remarkably increase the solubility of organic matter. It involves heating of the substance, usually to a temperature in the range of 150°C to 200°C. The pressures adjacent these temperatures are in the range of 5atm to 20atm [18].

This process produces a material with the solvent molecules that are more readily available and can be disintegrated in anaerobic digesters. The Porteous process and the Cambi process are examples of thermal hydrolysis concepts which were previously implemented on many sites worldwide [19, 20].

Comparing TH with other technologies shows that it is a continuously operated technology involving high and controlled pressure and temperature. Even though differences in system arrangement such as temperature, pressure, time, characteristic of waste and pilot scale have confirmed expected benefits in energy recovery from enhanced digestion performance.

The AD of organic material basically follows four steps. These include hydrolysis, acidogenesis, acetogenesis and methanogenesis [21]. The thermal hydrolysis process is expected to accelerate and anticipates the first step of the anaerobic process and break the long chain molecules into an easily digestible feed for anaerobic digestion. The second effect is considered by a transfer indissoluble products towards degradable compounds and also killing pathogens.

However, special effects of thermal hydrolyses depend on waste characteristics and instrument and duration of this situation on materials.

The use of AD in cooperation with thermal hydrolysis has shown to produce improved VS removal, COD removal and greater biogas production [22].

Thermal hydrolysis is a process that uses heat and pressure as a substitute for biological hydrolysis. There are several advantages to this process. First and the main advantage of the process is that thermal hydrolysis can be accomplished in approximately 30 minutes as opposed to multiple days for passing the hydrolysis step in digestion.

Second, it is known for its highly efficient for destroying pathogens. Third, and maybe most important of the thermal hydrolysis process is that the solids concentration in the feed can be increased to 10 to 12% solids. This is in contrast to a practical limit of 7% for unhydrolyzed solids and allows a more concentrated feed to take place.

2 Materials and Methods

Municipal solid waste has been applied as feeding materials in this research. Based on the composition of the feedstock and considering the purpose of the study which is tracking the effect of hydrolysis temperature and comparing with normal digestion, the same type of organic compounds should be used in both bioreactors. As a consequence, the waste materials provided from a specific source and placed in a plastic container sealed enough to prevent any emission. Due to criticality of quality and quantity of organic materials, measuring of Total Solid (TS) and Volatile Solid (VS) was carried out before starting the digestion process which the results are demonstrated in Table 1.

The experiments were done in two anaerobic glassy digesters with capacity of 5 liters. The cylindrical digesters are made up of double-glass layer which assure its sealing against any heat exchange with the environment. By this way, the temperature of inside of digester could be kept constant. The digesters have height of 30 cm and inner diameter of 15 cm. The process took place in the same environmental conditions (i.e., solid concentration of 8%, temperature of 35 ± 1°C) for both digesters. The materials stirred completely in order to provide homogenous feeds. The batch-load digester loaded and filled just at the first of the test and didn't reload by any input material during the test. A heating plate was used to preserve the temperature of the material inside the digester. The loading process was done using a magnetic stirrer in order that avoid over limitation of 8% solids as maximum allowed.

Stirring material in the digester not only stimulates microorganisms but also produce more gas subsequently. Digesters were investigated at time intervals of one month and averagely once in each 12 hr. In addition, rate of gas production was measured, recorded and finally discharged.

5kg of organic material with a moisture content of 54% was sampled and delivered to laboratory. The samples preserved in condition of 155°C and pressure of 5 atm for 30minutes and then was placed in plastic reservoir with appropriate lid and prepared for loading in digesters. Next, 1kg organic waste together with 2 lit of water added into the digesters. The first digester was loaded with normal organic material and second one loaded by thermal hydrolyzed organic waste.

By a magnetic stirrer and a heating plate at the bottom of the digester, digesters got warmed up and materials in digesters were mixed. Produced gas went gradually to storage tank with floating dome and fixed gas pressure equal to 1.07atm. Storage tank which consists of two tanks with diameters of 12cm and 13cm and height of 20cm placed upside down on each other and discarded 10cm water in bottom of the tank.

Table 1 Qualitative and quantitative characteristics of municipal solid waste

Parameter	Amount (Kg)	Percentage from Total
Average municipal solid waste	1000	100.00
Average home solid waste	830	83.00
Average biodegradable solid waste	538	53.79
Average total solid (TS)	66	6.59
Average total volatile solid (VS)	43	4.34

The displacement of the surface water between the two cylinders is a sign of the volume of gas production.

The upper tank was controlled by lightweight aluminum tubes which prevented its extra rising and collapse. Body of gas storage tank graded in centimeter and movement of water between two glass cylinders measured continuously in each 12 hr and then the output data registered.

During the experiments at 3 days, 7 days, 15 days and 21 days after start of the experiment, sampling was carried out from produced gas in outlet of each digester. Analysis of the samples was finally performed using gas-chromatography (GC-14B made by Shimadzu) apparatus. It should be noted that sampling did by special syringe of GC.

3 Results and Discussion

Doing described tests yielded the results that showed a decline RT in the first digester with normal organic waste and second digester with hydrolyzed organic waste to 20days and 15days, respectively (Figure 1–2).

Morgan-Sagastume et al., 2011, have reported that the efficiency of the process varied moderately with increase in temperature but the change in RT was seen to have a greater impact on the digester's performance [23]. Also the biogas produced from one kilogram of organic material in the first and second digester is 321.01 liters and 497.15 liters, respectively.

Some researchers have reported that thermal hydrolysis pretreatment [24–27] have been shown to have a positive effect on either the solubilization of organic material.

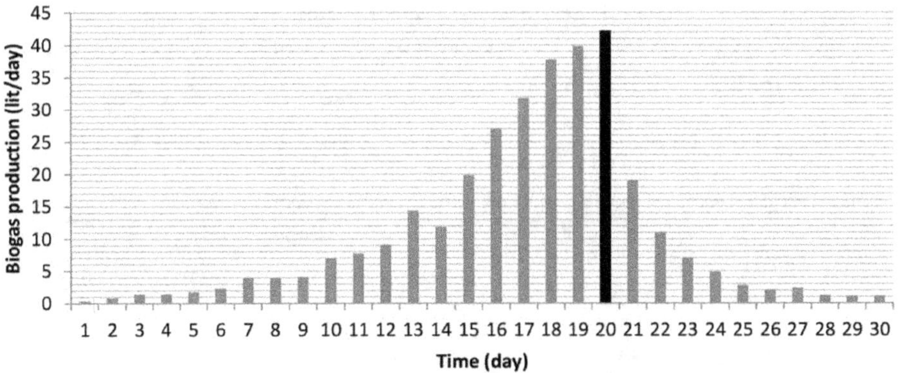

Figure 1 The biogas produced in the first digester during the 30-day trial.

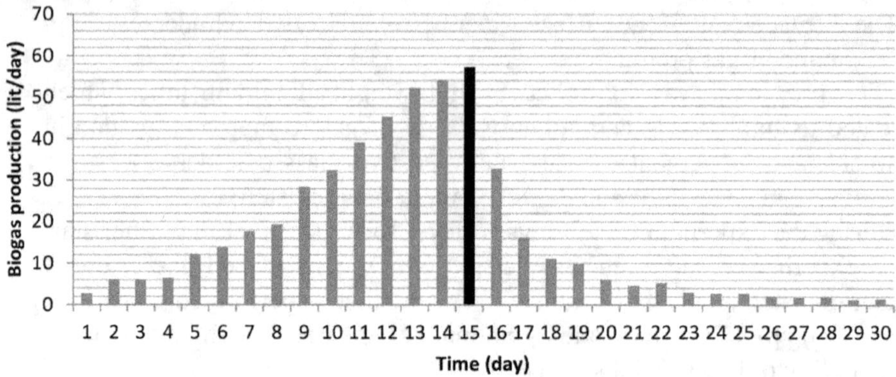

Figure 2 The biogas produced in the second digester during the 30-day trial.

According to the gas sampling in different days during the experiment and analyzing samples with GC results in Table 1 were obtained. Gas sampling has been taken from a plastic tube connected digester to gas storage tank for the lowest error analysis. The results obtained from this study have revealed that biogas produced in the second digester has higher and better quality than first digester and amount of methane up to 72% has been recorded (Table 2). High methane in biogas means an increase in the thermal value and better efficiency in digester.

Various authors have studied the effects of thermal hydrolysis treatment on the solid waste and sludge. While, all studies make known that thermal hydrolysis as a pretreatment has an optimistic impact on anaerobic digestion, it has also been observed that the hydrolysis conditions play an essential task in determining the efficiency of the process [28]. It had been reported that Thermal pretreatment in the temperature range from 100°C–180°C destroys cell walls and makes the proteins accessible for biological degradation [29].

Table 2 Characteristics of biogas in first and second digester

Day	First Digester			Second Digester		
	CH_4	CO_2	N_2	CH_4	CO_2	N_2
3	36%	45%	4%	49%	37%	4%
7	45%	40%	5%	62%	31%	4%
15	55%	28%	3%	72%	24%	2%
21	63%	22%	3%	68%	23%	3%

4 Conclusions

The greater than before demand for highly developed techniques in AD over the last few years has led to the employment of various pretreatment methods prior to AD to raise gas production. Experimental results indicate that the use of thermal hydrolysis as a pretreatment for anaerobic digester feed material reduced RT of 33.33% and this result means significantly reduces for the required land and facilities needed to digest defined amounts of organic waste and it will also seek to increase public acceptance of this technology. In the other hand, the biogas produced from one kilogram of organic material on the thermal hydrolysis is significantly increased 154.87%. Other achievements of this study increase quality of biogas produced in second digester which is loaded with hydrolyzed material, 72% was recorded for methane content in biogas produced in this digester which is considered an appropriate result.

In addition to all the good effects of thermal hydrolysis on the anaerobic digester efficiency, destroy pathogens in the waste caused by the high temperature pretreatment another one of the advantages of using this technology. As a result the digested material that is brought out of the digester can easily be used as fertilizer in agriculture.

References

[1] Tehran Organization of Waste Recycling and Composting, (2013).

[2] Hong, F., Tsuno, H., Hidaka, T., and Cheon, J. H.: Study on applicability and operation factor of thermophilic methane fermentation to garbage treatment in high concentration under.

[3] Park, Y., Hong, F., Cheon, J., Hidaka, T., and Tsuno, H.: Comparison of thermophilic anaerobic digestion characteristics between single-phase and two-phase systems for kitchen garbage treatment. J. Biosci. Bioeng., 105, 48–54 (2008).

[4] Van Lier, J. B., Tilche, A., Ahring, B. K., Macarie, H., Moletta, R., Dohanyos, M., Hulshoff Pol, L. W., Lens, P., Verstraete, W., 2001. New perspectives in anaerobic digestion. Water Science & Technology 43 (1), 1–18.

[5] Midilli A, Dincer I, Ay M. Green energy strategies for sustainable development. Energy Policy 2006;34(18):3623e33.

[6] Weizhang Zhong, Zhongzhi Zhang, Wei Qiao, Pengcheng Fu, Man Liu. Comparison of chemical and biological pretreatment of corn straw for

biogas production by anaerobic digestion. Renewable Energy 36 (2011) 1875e1879

[7] Kassam Z. A., Yerushalmi L., Guiot S. R. (2003) A Market Study on the Anaerobic Waste-water Treatment Systems. Water, Air & Soil Pollution 143:179-0-192.

[8] Borzacconi L, Lṕez I, Viñas M. Application of anaerobic digestion to the treatment of agroindustrial effluents in Latin America. Water Sci Technol 1995;32:105–11.

[9] Murto M, Bjornsson L, Mattiasson B. Impact of food industrial waste on anaerobic co-digestion of sewage sludge and pig manure. J Environ Manag 2004;70:101–7.

[10] Yen HW, Brune DE. Anaerobic co-digestion of algal sludge and waste paper to produce methane. Bioresour Technol 2007;98:130–4.

[11] Hartmann, H., Ahring, B. K., 2005. Anaerobic digestion of the organic fraction of municipal solid waste: influence of co-digestion with manure. Water Research 39 (8), 1543–1552.

[12] Davidsson, A., Gruvberger, C., Christensen, T. H., Hansen, T. L., Jansen, J. la C., 2007. Methane yield in the source-sorted organic fraction of municipal solid waste.Waste Management 27 (3), 406–414.

[13] Comino, E., Rosso, M., Riggio, V., 2009. Development of a pilot scale anaerobic digester for biogas production from cow manure and whey mix. Bioresource technology 100, 5072–5078.

[14] Komaki, H., Yamashita, M., Niwa, Y., Tanaka, Y., Kamiya, N., Ando, Y., Furuse, M., 1998. The effect of processing of Chlorella vulgaris: K-5 on in vitro and in vivo digestibility in rats. Anim. Feed Sci. Technol. 70, 363–366.

[15] Janczyk, P., Franke, H., Souffrant, W. B., 2007. Nutritional value of Chlorella vulgaris: effects of ultrasonication and electroporation on digestibility in rats. Anim. Feed Sci. Technol. 132, 163–169.

[16] Carrère, H., Dumas, C., Battimelli, A., Batstone, D. J., Delgenès, J. P., Steyer, J. P., Ferrer, I., 2010. Pretreatment methods to improve sludge anaerobic degradability: a review. J. Hazard. Mater. 183, 1–15.

[17] Carlsson, M., Lagerkvist, A., Morgan-Sagastume, F., 2012. The effects of substrate pre-treatment on anaerobic digestion systems: a review. Waste Manage. (Oxford) 32, 1634–1650.

[18] Abdoli, M. A., (2009), Municipal solid waste management, University of Tehran publication, 3rd edition.

[19] Kepp U., Machenbach I., Weisz N. and Solheim O. E. (2000) Enhanced stabilization of sewage sludge through thermal hydrolysis – three years

of experience with full scale plant. Water Science and Technology, 42 (9), 89–96.

[20] Neyens E. and Baeyens J. (2003) A review of thermal sludge pre-treatment processes to improve dewaterability. Journal of Hazardous Materials, 98(1), 51–67.

[21] Grady, C. P. Leslie, Daigger, Glen T., Lim, H. C., Biological Wastewater Treatment, Marcel Dekker, New York, 1999.

[22] Camacho, P., Ewert, W., Kopp, J., Panter, K., Perez-Elvira, S. I., Piat, E. Combined experiences of thermal hydrolysis and anaerobic digestion – latest thinking on thermal hydrolysis of secondary sludge only for optimum dewatering and digestion. Water Environment Federation, WEFTEC 2008.

[23] Morgan-Sagastume, F., Pratt, S., Karlsson, A., Cirne, D., Lant, P., Werker, A., 2011. Production of volatile fatty acids by fermentation of waste activated sludge pretreated in full-scale thermal hydrolysis plants. Bioresour. Technol. 102, 3089–3097.

[24] Bougrier, C., Delgenes, J. P., Carrere, H., 2008. Effects of thermal treatments on five different waste activated sludge samples solubilisation, physical properties and anaerobic digestion. Chemical Engineering Journal 139 (2), 236–244.

[25] Climent, M., Ferrer, I., Baeza, M. D., Artola, A., Vazquez, F., 2007. Effects of thermal and mechanical pretreatments of secondary sludge on biogas production under thermophilic conditions. Chemical Engineering Journal 133 (1–3), 335–342.

[26] Li, Y. Y., Noike, T., 1992. Upgrading of anaerobic-digestion of waste activated-sludge by thermal pretreatment. Water Science and Technology 26 (3–4), 857–866.

[27] Tanaka, S., Kobayashi, T., Kamiyama, K., Bildan, M. L. N. S., 1997. Effects of thermochemical pretreatment on the anaerobic digestion of waste activated sludge. Water Science and Technology 35 (8), 209–215.

[28] Wilson Christopher. A., Murthy, Sudhir N., Novak John T. Digestibility Study of Wastewater Sludge Treated by Thermal Hydrolysis. Residuals and Biosolids, pp. 374–386, 2008.

[29] Muller, J. A. Prospects and problems of sludge pre-treatment processes. Water Science and Technology, Vol. 44, No. 10, pp 121–128. 2001.

Biographies

L. Amiri obtained her master degree in Civil-Environmental Engineering from University of Tehran, in 2009. Currently she is doing her PhD in Environmental Engineering in University of Tehran. Her areas of interests include waste to energy techniques, renewable energy and biogas.

M. A. Abdoli obtained a PhD in Environmental Engineering, in 1982. Currently he is a full professor in University of Tehran. His research interests include solid waste management and waste to energy, biomass, renewable and sustainable energy.

N. Mehrdadi obtained a PhD in Environmental Engineering, in 1994. Currently he is a full professor in University of Tehran. His research interests include waste water treatment and waste to energy, biomass and renewable energy.

Prolonging the Lifetime of the Wireless Sensor Network Based on Blending of Genetic Algorithm and Ant Colony Optimization

Soumitra Das and Sanjeev Wagh

Dept. of Computer Engineering, Sathyabama University, Chennai, India
Dept. of Computer Engineering, KJ College of Engineering and Management Research, Pune, India
Corresponding author: Soumitra Das <soumitra_das@yahoo.com>

Received 21 January 2015; Accepted 12 February 2015;
Publication 19 March 2015

Abstract

The application of WSN has developed manifolds in the last few years. The lifetime of a sensor node is mainly focused around the battery fuelled gadgets. There are numerous strategies to save energy and one of the best viable techniques of saving energy was discovery of the multipath shortest route from source to destination for both data dissemination and data aggregation. In this paper, we propose a hybrid model for energy optimization focused on formation of Cluster and Cluster Head determination based on Genetic Algorithm. And then apply an Ant Colony Optimization algorithm to find the shortest path from source Cluster Head to destination Sink using multipath routing data transmission. This will help to obtain reliable communication in case of node failure by means of route refurbish mechanism. The fundamental objective is to retain maximum lifetime of the network throughout the data transmission phase in a proficient manner. The proposed algorithm was compared with Genetic Algorithm Based Energy Efficient Clusters [5] and Energy Efficiency Performance Improvements for Ant-Based Routing Algorithm [12] for energy efficiency. The effectiveness of the proposed algorithm is demonstrated by simulations.

Journal of Green Engineering, Vol. 4, 245–260.
doi: 10.13052/jge1904-4720.435

Keywords: ACO, Clustering, Data Aggregation, Energy Efficient Routing, EEABR, GABEEC, GA, WSN.

1 Introduction

Modern advancements in engineering, science and communication have motivated the development of low cost, low power, and tiny wireless sensor communication devices. However, the most challenging issue is the insufficient battery power of the sensor nodes. Designing of Wireless Sensor Network (WSN) algorithms needs to focus on the parameters to amplify the life span of the sensor nodes and in turn the entire network. Performance issues related to energy consumption are evaluated by Ehsan et al. [1]. Cluster based routing and data aggregation methods are the most admired techniques in WSNs for energy efficiency [2]. These research studies have demonstrated blending of, two research methods, Genetic Algorithm (GA) and Ant Colony Optimization (ACO) to reap the synergistic benefits in order to boost the energy efficiency of the sensor nodes. The proposed algorithm executes as follows:

1. Selection of Cluster Head (CH) and formation of clusters using GA.
2. Shortest path selection based on ACO algorithm.
3. Failure detection using Route Refurbish Mechanism and best shortest path selection using ACO for rerouting.

Section 2 related work briefly examines the existing work of GA and ACO protocols. Section 3 describes the proposed work, Section 4 illustrates the implementation and results. The need of future research and validation has been mentioned in the conclusion section.

2 Related Work

This section is divided in two parts and describes various methods of energy efficient routing algorithms. The first part focuses on GA based energy efficient selection of (CH) and forming of clusters and the second part explains about energy efficient ACO based shortest path discovery from source to the destination node.

2.1 GA based CH Selection and Cluster Formation Approaches

The demonstration of GA was first derived from Darwins theory of evolution [3]. Darwin suggested that the individual who is the fittest will continue to exist in the context of survival [4] [13] and others will eventually perish. Similar theory has been used by GA based CH selection. Here also, a fittest entity is considered to be the candidate for CH selection.

In one of the published article, [5], Selim et al. has projected a cluster based method which is comparable to Low Enery Adaptive Clustering Hierarchy (LEACH) and works in two phases: set-up phase and steady-state phase. In the set-up phase, predefined set of sensor nodes are chosen as CHs. Number of CHs and clusters are same as only one CH is related to each cluster. The member nodes opt to join nearest CH based on the distances. In the steady-state phase, all the nodes actually start the process of communicating with the CHs using the Time Division Multiple Access (TDMA) schedule. To complete a single round, the Cluster Head collects the data from the member nodes, fuses it, and packs the data into a single packet and sends to the Base Station (BS). After every round, energy of each CH is checked and if found fewer than the average energy of all the associate nodes, a new CH is selected based on the node that has the maximum residual energy and the old CH then becomes an associate node. Chromosomes stand for the network where CH is represented using 1 and associate node is represented using 0. Randomly the initial population is generated. The fitness of every node is calculated and based upon the fitness, the fittest chromosome is selected for applying crossover and mutation. The author has considered three parameters for selection of the fitness function. The parameters are Cluster distance C, which is the sum of distances from associate nodes of the CH and then from CH to the BS, the round at which the first node dies (R*fnd*), and the round at which the last node dies (R*lnd*).
The fitness function *(F)* is defined as

$$F = \sum_{i} (f_i * \omega_i) \forall f_i \in (R_f nd, R_l nd, C) \tag{1}$$

Thus, the projected protocol is better than the classical LEACH in terms of the number of alive nodes. Even though, it resulted in improvement of lifetime of the network along with achievement of energy efficient cluster formation and CH selection, it failed to change the clusters throughout the lifetime of the network. Also, only the CHs are rotated every time resulting into lack of efficiency and throughput of the network.

Furthermore, in their paper [6], Abbas et al. has proposed a method where in the fitness function calculation depends upon difference of energy of chromosomes in the current and the previous round. Chromosome with least difference gets selected. The proposed algorithm initializes the network, and eventually each and every sensor node sends its positions to its corresponding neighbour nodes. It calculates the fuzzy parameter 'Chance' by considering fuzzy descriptors such as energy, density and the centrality of the nodes. The nodes that have the higher 'Chance' than their neighbours will be selected as the candidate for CHs. The BS applies a GA based on chaotic and selects the CHs from the candidates pool list and then the new CH broadcasts to the network and the member nodes which are nearest to the CH will then join and form clusters. This proposed method is successful in increasing the lifetime of the network. The only disadvantage is that while calculating the fitness, only energy is taken into account. The implied hypothesis is that if the distance was considered, the results may have been better.

In this paper [4], Sanjeev et al. has projected a GA for optimizing the sensor nodes energy consumption with clustering techniques. A multi-objective algorithm enabled generation of an optimal number of sensors-clusters with CHs. This approach reduced the expenditure of transmission. The author has used the theory of multi-objective to reduce energy utilization and maximizing the coverage area in the WSN by considering the regular genetic process, linear ranking technique along with selection pressure. The fitness for every sensor node is calculated using ratio of Minskowsk's distance between all solution pairs in the normalized objective space and the niche count for each solution. The authors inferred that the line of attack involves a new means for cluster formation which ultimately lengthens the lifetime of the network through equally distributed clustering.

2.2 ACO for Shortest Path Selection Approaches

ACO based energy efficient routing protocols have the ability to select the shortest path among the possible paths from the source to destination.

In this particular literature [7], Camilo et al. has shown a better version of the ant based routing in WSN, the Energy Efficiency Performance Improvements for Ant-Based Routing Algorithm (EEABR). This method considers the nodes in terms of energy level and distance of the path navigated by the ants.

The recommendation of the authors was that in the basic ant algorithm, the forward ants are propelled to no explicit target node. This indicates that sensor nodes need to converse with each other and also, the routing tables of each node should hold the identification of all the sensor nodes in the neighbourhood and the succeeding stages of pheromone trail. In their work, success was recorded in energy savings to a great extent, but encountered ambiguity in the mobility and dynamic setting as lot of control traffic was produced hence consuming a lot of energy reducing reliability.

In this paper [8], Liu et al. has projected a routing method, taking into account the ACO based algorithm. The ant uses the routing mechanism based on the angle of deflection, energy and distance as routing aspects. Although the convergence rate of this particular algorithm is admirable, the algorithm fails to make use of redundancy of data. Thus, the disadvantage is the data correlation. However, the energy expenditure of the communication is vast when a lot of sources are present in the network.

In this paper [9], Ren et al. has projected a multipath routing protocol based on ant colony system, which extends the network life span. Even though the algorithm balances the energy utilization among nodes by multipath, it does not take into consideration the influence of the minimum energy node on multiple paths.

The authors of this paper [10], Nikolidakis et al. have demonstrated a new protocol called equalized CH election routing. This protocol pursues energy saving through balanced clustering. Energy efficient routing in WSNs through balanced clustering algorithm models the network as a linear system using the Gaussian elimination algorithm. It then calculates the mixture of nodes that are probable CHs in order to amplify the network life span. This protocol is definitely proficient in terms of network life span when evaluated against other well known protocols.

3 The Proposed Approach

The operation of the proposed approach is alienated in three stages to do the entire mission of sending the desired data to the sink through CH. The proposed three phases: phase 1, the selection of CH and formation of clusters using GA, phase 2, ACO algorithm for finding and selection of shortest path and the final phase 3, will be responsible for routing using route refurbish mechanism. The proposed system model is shown in Figure 1.

3.1 Cluster Formation Using GA

The objective of the cluster formation is to prolong the network lifetime by means of clustering along with maximizing the working time of the cluster and minimizing the energy consumption in the cluster. The conventional hierarchical clustering algorithm does not provide accurate formation of

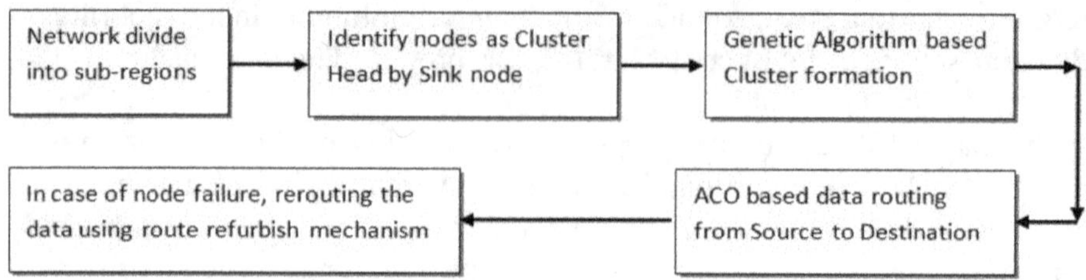

Figure 1 Proposed system model.

clusters and selection of CHs. If the cluster is reconstructed every time, then it will unnecessarily use energy reducing the data aggregation efficiency.

In order to overcome this, our phase 1 adopts a GA based dynamic clustering technique. GA first finds the optimum number of CHs based on residual energy, radio frequency signal strength and centrality of the nodes. Once the CH is chosen, then each member sensor node will join the nearest CH based on the Euclidean distance between CH and the member node. For a cluster with 'm' static member nodes, cluster distance (cd) is defined as

$$f(cd) = \sum_{i=0}^{m} dist(i, ch) \tag{2}$$

Where, *dist* (*i, ch*): is the Euclidean distance between ith node and the CH. The energy function of a wireless sensor node is equivalent to the fitness function of GA based on the negligible consumed energy from network nodes in every generation. Here, it maintains a population of individuals which is called a chromosome. Every chromosome is evaluated by an utility known as fitness function. Next fresh population is generated from the current one through selection, crossover and mutation described as follows:

Selection Mechanism: The purpose is to select more healthy individuals (parents) for crossover and mutation.

Crossover: It causes swap over of the genetic materials among parents to form offspring.

Mutations: It incorporates new genetic resources in the offspring. It is possible that a regular node may become CH and a CH may become a regular node.

Finally, after crossover and mutation are over, the BS selects the chromosomes which have the networks smallest amount of energy difference in percentage of the previous round and introduces the existing nodes in the network as CH and other nodes connect to the nearest CH.

3.2 Pseudo Code for Cluster Formation

1. Initialize the population P (t) at t = 0, where t = time.
2. Compute the fitness P (t) based on residual energy, radio frequency signal strength and centrality.
3. Increment t by 1.
4. If fitness is less than the required threshold, then terminate.
5. Select P (t) from the P (t-1) list.
6. Crossover P (t).
7. Mutate P (t)
8. Repeat step 2 to 7 till all the fitness of the nodes is tested.
9. Then the BS selects the CH based on the fitness of the node.
10. Introduce CH to all nodes in the network.
11. Each sensor node joins to the nearest CH based on Euclidean distance i.e. shortest distance.
12. Each sensor node transmits data to the CH with a single hop transmission only.
13. After all data has been received, CH aggregates all the data and then transmits it to the BS through multiple hops using ACO based shortest path algorithm.

3.3 Shortest Path Routing Based on ACO

The basic concept was derived from the ant colony where the ants are placed initially in the Source node (Sn) with the task of finding out paths through the in-network nodes to the destination node (Dn). An ant going from the Sn to Dn collects information about the quality of the path, and uses this information to update the pheromone of the best paths in the pheromone table [11].

The proposed method uses ACO algorithms to find multiple routes from Sn to Dn, after the cluster formation process is completed. As when an event occurs, the member nodes send the gathered data to the respective CH. Then, to route the data from Sn to Dn, the CH sends a route establishment message to the neighbour's CH node. When a neighbour CH node receives the route establishment message, it checks for the next neighbour CH node and this process continues till it gets the Dn and eventually a hop tree is formed.

This process is repeated multiple times from Sn to Dn to get multiple paths. Having multiple paths can maximize the network lifetime by transferring the data from Sn to Dn through the shortest path and also maximizes the reliability. The advantage is due to multiple routes because if one route fails, the data could be re-routed through the alternate route having minimum distance from Sn to Dn. The objective is to maximize the network lifetime and minimize the energy consumption of the nodes.

3.4 Pseudo Code for Shortest Route Selection

1. Initialize the parameters
2. Initialize array heuristic
3. Initialize the pheromone matrix.
4. Stop when conditions are not satisfied
5. Build solutions
6. Apply local search
7. Update pheromone
8. End
9. View best solution
10. Stop

3.5 Pheromone Table Update

The responsibility of the pheromone table is to maintain the information collected by the Forward Ant (FA). Every node maintains a table which records the amount of pheromone on each neighbour lane. The node has a distinct pheromone scent, and the table is in the shape of a matrix as shown in Figure 2.

The rows represent the destinations and columns represent the neighbours. An entry in the pheromone table is indicated by Pn,d where 'n' is the neighbour index and 'd' stands for the destination index. The contents in the pheromone table are used to compute the selecting probabilities of every neighbour. The routing table is updated at each node by the following equation

$$\tau(x,y) = (1-\rho) * \tau(x,y) + \left[\frac{\Delta\tau}{\phi Bwl}\right] \tag{3}$$

It is important to note that all neighbours are probable destinations in the route selection method of routing.

Neighbours of Node 'A' →	Node 1	...	Node m
Destinations	$P_{n,d}$	...	$P_{n,d}$
Node 1 ↓	$P_{n,d}$	...	$P_{n,d}$
⋮	⋮	...	⋮
Node m	$P_{n,d}$	...	$P_{n,d}$

Figure 2 Pheromone table of node A.

Legend	Explanation
ϕ	Coefficient
Bwl	distance travelled by backward ant l
ρ	is coefficient such that $(1-\rho)$ *represents the* evaporation of pheromone trails since the last time $\tau(x, y)$, *when updated*

3.6 Route Refurbish Mechanism

Route maintenance plays a very significant role in WSN's as the network keeps changing dynamically and the routes found good during discovery may turn to be bad. Possible cause of failure includes low energy, physical destruction and communication blockage. The route created to send the data toward the sink node is unique and efficient, any failure in one of its nodes cause disruption, preventing the delivery of aggregated data to the neighbour node. Our algorithm offers a route refurbish mechanism by finding the goodness of a route regularly and updates the pheromone counts for the different routes at the source nodes. To accomplish this, when a destination node receives a packet, it sends an acknowledgement message to the source informing the well being of that route.

As seen in the Figure 3 the data is routed from source node to the destination node through route 1, but when it found that the node is not responding due to failure of the node i.e. no acknowledgement is sent back to the source. The source initiates the route refurbish algorithm and selects another route based on the shortest path and data is transmitted through route 2 to the destination.

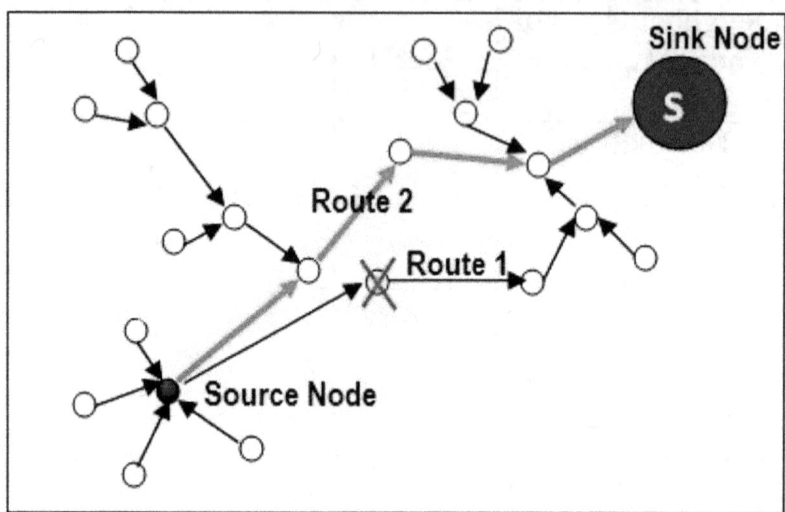

Figure 3 Route refurbished mechanism when route fails.

4 Implementation and Results

The proposed algorithm was simulated using MATLAB R2009b and compared with Genetic Algorithm Based Energy Efficient Clusters (GABEEC) [5] and Energy Efficiency Performance Improvements for Ant-Based Routing Algorithm (EEABR) [12]. Figure 4 and Figure 5 shows the simulation parameters along with their values.

Parameter	Value
No. of Nodes	100
Initial Energy	100 Joules
Area	100m X 100m
Receive Power	22.2MW
Transmit Power	31.2MW
Data Packet Size	40 bytes
Base Station Location	120,50
Sleep Power	0.0006mW
Noise bandwidth	30KHz

Figure 4 WSN parameters.

Parameter	Value
Max Generations	500
Population Size	200
Length of chromosome	100
Mutation Rate	0.007
Crossover Rate	0.7

Figure 5 GA parameters.

The proposed methodology has used the random deployment model for the WSN topology set-up. The BS is placed in the location (120, 50) away from the sensor field. We have compared our results with GABEEC [5] and EEABR [12] protocols. Energy is one of the major issues in wireless sensor network.

Figure 6 shows the energy consumption of network with respect to time. From Figure 7 clearly depicts that our proposed algorithm is better in terms of energy consumption by almost 72 percent as compared to other two protocols.

Figure 8 presents the obtained energy value in each round with respect to the number of rounds. Also, from Figure 9 it is evident that our proposed algorithm is superior in terms of energy value (left over network energy) by 9 percent as compared to GABEEC and EEABR protocols.

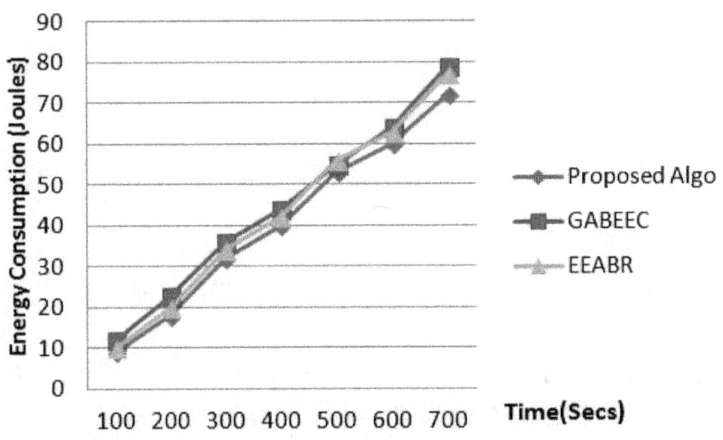

Figure 6 Energy consumption with respect to Time.

	Energy Consumption (Joules)		
Time (Secs)	EEABR	GABEEC	Proposed Algorithm
100	10%	12%	9%
500	56%	55%	53%
700	77%	79%	72%

Figure 7 Comparison of different protocols for energy consumption.

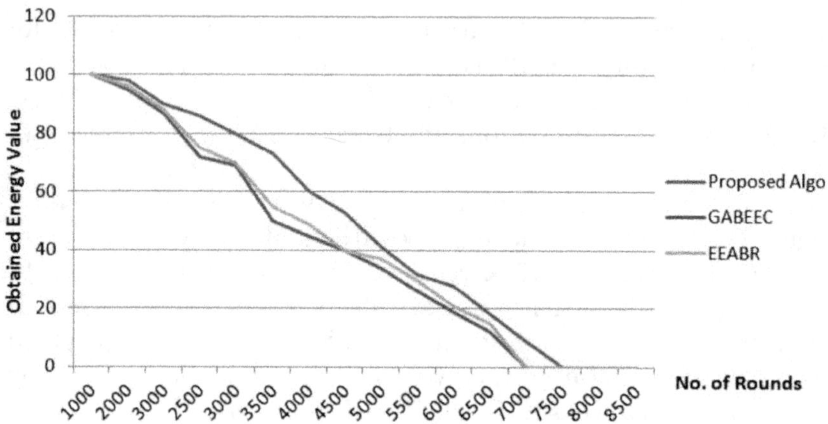

Figure 8 Energy comparison in each round.

	Energy Value		
No. of Rounds	EEABR	GABEEC	Proposed Algorithm
2000	96%	95%	98%
5000	37%	34%	41%
7500	0%	0%	9%

Figure 9 Comparison of different protocols for energy value.

Figure 10 shows the number of alive nodes with respect to time that is the lifetime of the network. From Figure 11, it is apparent that our proposed algorithm is also better in terms of alive nodes in the network as against the GABEEC and EEABR protocols.

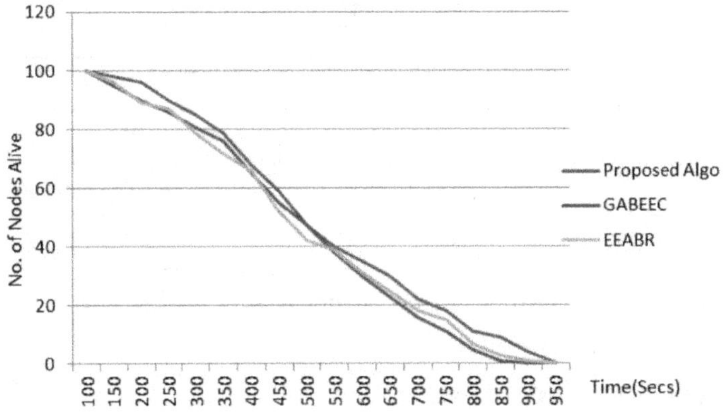

Figure 10 Network lifetime.

	Number of Alive Nodes		
Time (Secs)	EEABR	GABEEC	Proposed Algorithm
200	89	90	96
500	42	47	47
900	1	0	4

Figure 11 Comparison of different protocols for number of alive nodes.

Figure 12 illustrates the throughput of the network. Lastly from Figure 13 it is noticeable that our proposed algorithm is an improved throughput by approximately 32 percent as compared to GABEEC and EEABR protocols.

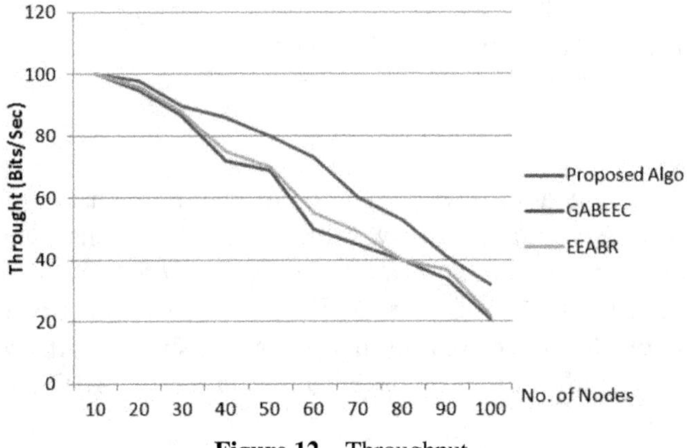

Figure 12 Throughput.

	Throughput (Bits/Sec)		
No. of Nodes	EEABR	GABEEC	Proposed Algorithm
40	75	72	86
80	40	40	53
100	22	21	32

Figure 13 Comparison of different protocols for throughput.

5 Conclusion and Future work

In this paper, a novel algorithm has been proposed based on the blending of Genetic Algorithm and Ant Colony Optimization for the dynamic formation of clusters according to the position of the nodes and Cluster Head and the types of values, it is sensing. Cluster information is broadcast to each of the nodes and routing is formed. The results of our simulation study indicate that the proposed method conserves more energy than existing established methods. In future work , the focus will be to improve the throughput and network lifetime along with addition of new parameters like packet delivery ratio, packet loss ratio and end to end delay.

6 Acknowledgement

We would like to express our gratitude to Dr. B.P. Patil and Dr. Santosh S. Sonavane for their mindful and inventive remarks, all through the readiness of the paper. Last yet not the least, all my colleagues, family members, friends and my wife Dr. Soma for their colossal support.

References

[1] Ehsan S. and Hamdaoui B. A Survey on energy efficient routing techniques with QoS assurances for wireless multimedia sensor networks. *Communications Surveys Tutorials, IEEE,*14 (2): 265–278, 2012.

[2] Singh S. K, Singh M. P Singh D. K. A survey of energy efficient hierarchical cluster based routing in wireless sensor networks. *International Journal of Advanced Networking and Application,*2 (02): 570–580, 2010.

[3] Jianming Zhang, Yaping Lin, Cuihong, Zhou, Jingcheng Ouyang. Optimal Model for Energy-Efficient Clustering in Wireless Sensor Networks Using Global Simulated Annealing Genetic Algorithm. *International Symposium on Intelligent Information Technology Application Workshops*, 2008.

[4] Sanjeev Wagh and Ramjee Prasad. Heuristic Clustering for Wireless Sensor Networks using Genetic Approach *International Journal of Wireless and Mobile Networking* (*IJWAMN*), Vol.1, No.1, 51–62, 2013

[5] Selim Bayrakli and Senol Zafer Erdogan. Genetic algorithm based energy efficient clusters (GABEEC) in wireless sensor networks *The 3rd International Conferernce on Ambient Systems, Networks and Technologies,* 247–254, 2012.

[6] Abbas Karimi, S. M. Abedini, Faraneh Zarafshan and S. A.R Al-Haddad. Cluster Head Selection Using Fuzzy Logic and Chaotic Based Genetic Algorithm in Wireless Sensor Network *J. Basic. Appl. Sci. Res,* 3 (4): 694–703, 2013.

[7] T. C. Camilo, C. Carreto, J. S. Silva and F. Boavida. An energy efficient ant based routing algorithm for wireless sensor networks, *in proceedings of 5th International workshop on ant colony optimization and swarm intelligence,* Brussels, Belgium, 49–59, 2006.

[8] Liu. Y, Zhu. H, Xu. K and Jia. Y. A routing strategy based on ant algorithm for WSN, *In proceedings of the 3rd International Conference on natural computation,* Haikou, Hainan, China, 685–689, 2007.

[9] Ren X, Liang H and Wang Y. Multipath routing based on ant colony system in wireless sensor networks *In Proceedings of international conference on computer science and software engineering*, Wuhan, Hubei, China, 202–205, 2008.

[10] Nikolidakis S. A, Kandris D, Vergados D. D and Douligeris C. Energy efficient routing in wireless sensor networks through balanced clustering algorithms. *ddd*, 29–42, 2013.

[11] Farooq M and Caro GA. Routing Protocols for next-generation networks inspired by collective behaviours of insect societies An overview. *Swarm Intelligence*, 101–106, 2008.

[12] Zungeru, A. M., Seng, K. P., Ang, L. M., and Chong Chia, W. Energy Efficiency Performance Improvements for Ant-Based Routing Algorithm (EEABR) *in Wireless Sensor Networks. Journal of Sensors*, 2013.

[13] Nivedita B Nimbalkar and Soumitra S Das. A Survey on Cluster Head Selection Techniques *in Multidisciplinary Journal of Research in Engineering and Technology*,Vol.1 Issue 1, 01–05, 2014.

Biographies

S. Das received his Bachelor degree in Computer Engineering from North Maharashtra University, Jalgaon, Maharashtra, India and Master degree in Computer Engineering from University of Pune, Pune, Maharashtra, India. Currently, he is PhD researcher at Sathyabama University, Chennai, India. His research interest includes Computer Networks, Wireless Sensor Networks, etc. He is a members of IEEE, CSI, LMISTE, IACSIT and IAENG. He is also an active reviewer of various conferences and journals.

S. Wagh received his Postdoctoral from Center for TeleInfrastruktur (CITF), Aalborg University (AAU), Denmark; PhD in Computer Science and Engineering from the SRT Marathwada University, Nanded and Master degree from University of Pune, India. He is a member of IEEE, ACM, FIETE, FIE, AND LMISTE. He is an active Steering committee member for the various International conferences and workshops; a technical program committee member and a reviewer for numerous top- quality conferences and journals in Wireless Networking and Simulation.

www.ingramcontent.com/pod-product-compliance
Lightning Source LLC
LaVergne TN
LVHW012333060326
832902LV00011B/1867

* 9 7 8 8 7 9 3 2 3 7 5 8 2 *